The True Story of Modern Cosmology

Emilio Elizalde

The True Story of Modern Cosmology

Origins, Main Actors and Breakthroughs

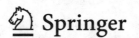 Springer

Emilio Elizalde
Valldoreix, Barcelona, Spain

ISBN 978-3-030-80656-9   ISBN 978-3-030-80654-5   (eBook)
https://doi.org/10.1007/978-3-030-80654-5

Based on a translation of the book *Cosmología moderna desde sus orígenes* © Los Libros de la Catarata, Madrid (2020)

Cover Image Credit:
Front Cover: Westerlund 2. Credit. NASA, ESA, the Hubble Heritage Team (STScI/AURA), A. Nota (ESA/STScI), and the Westerlund 2 Science Team
Back Cover: A Deep Look into a Dark Sky. Credit ESO

This Springer imprint is published by the registered company Springer Nature Switzerland AG
The registered company address is: Gewerbestrasse 11, 6330 Cham, Switzerland

*To Maria Carme
and our precious family
with endless love*

Preface

Having left behind a substantial part of my scientific life, initiated my emeritus period, and feeling an imminent need to respond without further delay to the insistent requests of an ever-growing number of colleagues—to write down '*those nice talks of yours*'—I eventually found in the COVID-19 pandemic the golden opportunity I was looking for. Since the middle of March 2020, I have been able to use most of my time to compile and bring together what I have learned in those many years of work at the frontier of knowledge, to compose an all-inclusive story, to find its deepest meanings, and to finally set everything in the plainest, most intelligible way I can. During this time, I could not stop writing, correcting, and rewriting again, night and day. The result is this book, which is not very thick, actually.

And it is at that point, when you feel trapped in the middle of a project like the present—having committed yourself to the search for the origins of '*everything*'—that you soon realize the necessity to explain, along the way, a number of not-so-simple concepts. Some are particularly difficult to grasp—and this with almost no equations, as required for a popular book. Certainly, everybody can find all these definitions—although not necessarily very well explained, in fact—in the abundant outreach bibliography, which grows endlessly day by day. And very soon, it becomes clear that what we are trying to do here would best be left to specialists, the really competent fellows in such complex matters. But those who really know all the intricacies of the subject rarely have time to devote to such a commendable purpose, since they spend all the time they have on their everyday work at the forefront of knowledge.

Achieving wisdom, in whatever subject, is a noble goal and one that can often lead to addiction. Particularly in the case at hand, in the study of the Universe, our extended home as intelligent beings. We want to know exactly how it is, how it was in the past, and what it will become of it. How did everything start and how will it eventually end, even though we shall never actually see this to happen. These are indeed very profound queries, which arise from the innermost part of our being, from our soul, in fact. We have learned a lot since humanity came into existence. And the truth is, the more we learn, the more we find we need to know. This is an indisputable and unavoidable certainty; a dilemma, if you like, which is going to re-appear in various situations throughout the book, but which in no way should hinder or condition our search for knowledge.

Here, I will restrict myself mainly to a quite specific issue: to identify the moment when cosmology—which is probably the oldest discipline of knowledge, the first to come into being at the dawn of human consciousness—finally became a true science. This did not happen overnight through the enactment of some definitive laws of nature or due to some amazing astronomical discovery. It actually occurred through a long, slow process, which took up much of the twentieth century and involved a great many astronomical researchers, serious (but also sometimes bombastic) cosmologists, and theorists of fundamental physics—with their great breakthroughs and their gross (and, with hindsight, quite incomprehensible) errors, as we will soon discover.

In writing the book, I could only mention a few of those who contributed to this extraordinary feat. I have tried to stick to a chronological account, whenever possible; the unavoidable deviations I will always justify and try to keep to a minimum. And I have highlighted in various places the small but very important lessons that we can learn from this story. It is now easier for us to identify them clearly, from the standpoint of our present understanding, given the very broad perspective we now have of the exceptional discoveries made over the last hundred years. These lessons can be of enormous use to us in understanding the present and in trying to determine the possible future of our beloved universe.

Valldoreix, Spain
April 2021

Prof. Dr. Prof. H.C. mult. Emilio Elizalde
National Research Council of Spain,
ICE-CSIC, IEEC
TUSUR University

Acknowledgments

Having written this book is already a way of showing my gratitude to all those colleagues and friends who, during my scientific career, have shared with me, with much patience and generosity, their precious knowledge, acquired after many years of hard work. It was precisely these colleagues (some of them quite prestigious) who, after listening to my lectures, urged me to write this story in detail. I do not give their names here, as they will be cited throughout the book, although I feel I should mention Enrique Gaztañaga, for a careful reading of the manuscript and important observations, my Springer Nature editor, Angela Lahee, for wonderful words of praise upon receiving my first draft, and for her constant encouragement and dedication, Rüdiger Vaas for a very careful review, important comments, and numerous relevant suggestions, and Stephen Lyle, for such a professional improvement of my primitive English in the last version of the book. I also thank the Spanish Royal Physics Society and, in special, his president and friend Prof. Adolfo Azcárraga who asked me to write the first version of this book, in Spanish, and helped me to get it published. To finish, I appreciate the interchange of ideas with a number of loyal friends and the love and understanding of my wonderful family.

Contents

1

Introduction: The Awakening of Cosmic Consciousness

If our true purpose were to really try to find the actual origins of cosmology—that is, to go back to the earliest attempts our ancestors made to know about the world around them—we would definitely need to go back very far in time, to the very dawn of human pre-prehistory. That would in itself be a journey as exciting, if not more so, than the one that aimed to find the sources of the Nile river. And it should be mentioned that, even today, the search for those sources—a secret that remained hidden for millennia—continues to unleash the passions of enthusiasts who study the history of exploration. The ultimate solution to this riddle was described by Sir Harry Johnston as the greatest geographical discovery after the discovery of America. In my opinion, it is only with feats like these that one can compare our present task of answering the question: *When did the awakening of cosmic consciousness actually occur?* The issue is indeed complex and tangled; unfortunately, it lies beyond the scope and pretensions of this little book. In it I will treat, in essence, what is known as modern cosmology—whose origin I place, for reasons that will be explained later, in the year 1912. It also coincides precisely with the moment when cosmology could finally become a true science. This occurred when it became able to make use of the most advanced theories of physics, which had just been put together after the great scientific revolutions that took place during the first third of the twentieth century. Specifically, these fundamental laws of *the* science *par excellence* allowed scholars to describe, and in principle understand, the current structure, evolution, and behavior of the Universe as a whole. And these same

© The Author(s), under exclusive license to Springer Nature
Switzerland AG 2021
E. Elizalde, *The True Story of Modern Cosmology*,
https://doi.org/10.1007/978-3-030-80654-5_1

Fig. 1.1 Staring in awe at the night sky. The Milky way with the rho ophiuchi star system from the top of El Teide in Tenerife, on a perfect summer's night. Author: AstroAnthony. Date: 12 June 2018, 01:26:13. CC BY-SA 4.0

laws, taken to the extremes—even though they are not in fact valid at those extremes—also allow us to get a fairly plausible idea about how and when the origin of the Cosmos took place and about what, predictably, will be its future and its end. But we will discuss all that later, in the following chapters. The rest of this first, introductory one will be devoted to the brief account, initiated above, of the origins and subsequent evolution of cosmology during the past centuries.

The Universe contains absolutely everything—it is the All (das *Weltall*, in German)—at any level, at any scale, at any time. Thus, our knowledge of it can be considered inextricably linked to the very awakening of the thoughts, reasoning, and dreams, in the newly formed mind of the primitive *Homo sapiens sapiens*.[1] It is not difficult to imagine that, looking ecstatically at the night sky and wondering in awe about what is out there, it is something that has happened since time immemorial. Often is said that a picture is worth a thousand words. And contemplating pictures such as those in Figs. 1.1, 1.2, 1.3 and 1.4 it is not difficult to conclude that this event was very likely simultaneous with the awakening of consciousness itself. And it is certainly

---

[1] "*Do Androids Dream of Electric Sheep?*" Philip K. Dick (1968). The reader should reflect for a moment on the profound questions and ideas raised in that little book.

Fig. 1.2 The night sky at the eastern tip of Scout Key in April 2018. Author: Viktorwills. CC BY-SA 4.0

Fig. 1.3 The night sky on the winding road connecting the ALMA operation support facility at 3000 m altitude to the array operation site (5000 m high) passes an area between 3500 and 3800 m dominated by large cacti (Echinopsis Atacamensis). These cacti grow on average 1 cm per year, and reach heights of up to 9 m. Stephane Guisard captured the beautiful sky above this unique location in the Chilean Atacama Desert. The Milky Way is seen in all its glory, as well as, in the lower right, the Large Magellanic Cloud. *Source* http://www.eso.org/public/images/milky-way-cactus/. ESO/S. Guisard. Date: 4 April 2011. CC BY-SA 4.0

Fig. 1.4 From time immemorial, through the megaliths of Argimusco, the visitor can admire the Milky Way with the naked eye. A natural spectacle that frames the Sicilian Stonehenge, in the Bosco di Malabotta, an oriented nature reserve. Author: Vincenzo Miconi. Date: 28 June 2019. CC BY-SA 4.0

possible that this occurred long before the same awareness had fully hatched in *sapiens*. We should not forget that our ancestors lived in the savannah and, in the evening, they would have had a spectacular celestial vault as their roof (much more impressive than the one we can now barely see from our polluted cities). I am not a specialist in this matter, although I confess to be really passionate about it. And, in order to write this first, introductory chapter, I decided to do a little research on this point, which I summarize here. I should confess the attraction I already felt in my young student years for the spectacular discoveries of the Leakey family [1] at the archaeological site of Olduvai Gorge, in the Great Rift Valley in Tanzania. Reading those fascinating books about Lucy and everything around her carried me back to the very origins of humanity.

To start with, in order to come to understand the foundations of the thinking capabilities of primitive man, we must begin with one of the most confusing principles of human thought: the *origins of consciousness* itself. In other words, when did we start to have knowledge of ourselves, to be creative and aware? Scientists agree on the most important steps in early human evolution. Our first ancestors appeared between five and seven million years ago, probably when some similar creatures, in Africa, began habitually to walk on two legs. Two and a half million years ago, they started using raw stone tools, without yet working them. And half a million years later, some of them spread from Africa to Asia and Europe. One of the first known humans was *Homo habilis*, who lived between 2.4 and 1.4 million years ago in East and Southern Africa. Others include *Homo rudolfensis*, who lived in East Africa between 1.9 and 1.8 million years ago (the name comes from its discovery east of Rudolph, Kenya); and *Homo erectus*, who ranged from southern Africa to present-day China and Indonesia, approximately 1.89 million to 110,000 years ago. In addition to these early humans, researchers have found evidence of an unknown *superarchaic* group, which separated from other humans in Africa about two million years ago. These superarchaic humans mated with the ancestors of Neanderthals and Denisovans [2]. This is the oldest known case of different human groups mating with each other, something that was already known to have happened, but much later.

The tools of the Oldovian, as the first human stone-working industry is called—born in Africa 2.7 million years ago, and named after the Olduvai Gorge mentioned earlier—are among the first to be used by our ancestors. And it has become clear that there is a stark contrast with the tools of the Acheulian, which began 1.8 million years ago and extended up to 100,000 years ago. The fact that much more advanced forms of cognition are required to create Acheulian hand axes means that the date of this type of cognition, closer to human, can be traced back to at least 1.8 million years ago. And, surprisingly, it has been found that the parts of the brain that are used to make these tools are precisely the same as those we dedicate to much more modern activities, such as playing the piano. After 800,000 years of making simple Oldovian tools, the first humans began making Acheulian axes about 1.8 million years ago. Some studies hypothesize that this advance led to a profound evolutionary change in the cognitive and linguistic abilities of the hominid, using a neuroarchaeological approach to support this suggestion.

1.1 The First Conscious Knowledge Was of a Geometric Nature

Thus, at some point about seven or eight hundred thousand years ago, a striking sensitivity to geometry and the perception of patterns allowed humans to begin making very refined Acheulian tools, all of them endowed with a certain symmetry. It is very unlikely that this would have been possible without an implicit knowledge of geometry, already embedded in their brains. Although some researchers still believe that the first marks were symbolic rather than aesthetic, it now seems increasingly likely that this was not the case and that this paradigm will have to be changed. Patterns have also been found engraved on shells made by *Homo erectus* about 540,000 years ago, and an intriguing observation of these ancient marks is that they all have grids, angles, and repetitive lines, and that they closely resemble each other over an immensely long period of time (Figs. 1.5, 1.6 and 1.7). If marks were symbolic, we would expect to see much more variation in them, in space and time, as we see in modern writing systems. On the contrary, this persistence in the designs indicates quite clearly that these symbols are more likely to relate to geometry or mathematics, which remain forever essentially unchanged. On the basis of my modest knowledge, I would conjecture that geometry was incorporated into human consciousness many thousands of years before any form of language, and long before *Homo sapiens* appeared.

Fig. 1.5 Early marks. Top, left to right: Trinil shell, Blombos engravings (two examples). Middle: South Africa on ostrich eggshell. Bottom: Gibraltar by Neanderthals on rock surface. Derek Hodgson, The Conversation, 2019. CC BY-ND 4.0

Fig. 1.6 Image of the Blombos Cave silcrete flake L13 displaying the lines that form a cross-hatched pattern. Image credit, C. Foster. From Christopher S. Henshilwood, Francesco d'Errico, Karen L. van Niekerk, Laure Dayet, Alain Queffelec and Luca Pollarolo, An abstract drawing from the 73,000-year-old levels at Blombos Cave, South Africa, Nature 562, 115–118 (2018). Copyright Springer Nature

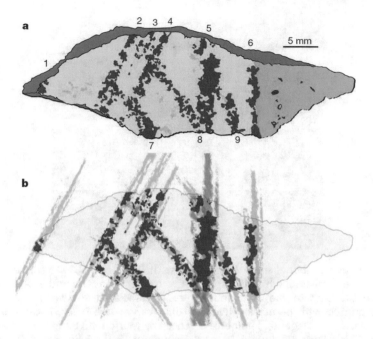

Fig. 1.7 Renditions of L13. **a** Tracing with the drawn red lines numbered, the calcite patches shown in orange, and the ochre residues in dark red. The ground surface is colored light grey, the darker grey area indicates a flake scar, and the darkest grey indicates the breakage fractures after L13 became detached from the originally larger grindstone. **b** Schematic of lines extending beyond the outline of the present flake. From Christopher S. Henshilwood, Francesco d'Errico, Karen L. van Niekerk, Laure Dayet, Alain Queffelec and Luca Pollarolo, An abstract drawing from the 73,000-year-old levels at Blombos Cave, South Africa, Nature 562, 115–118 (2018). Copyright Springer Nature

In 1999, anthropologist Chris Henshilwood made an intriguing discovery at a site in Blombos, on the east coast of South Africa (Fig. 1.6). He had been digging a prehistoric cave for more than a decade, containing several well-made artifacts, bone points, and spearheads dating back 73,000 years, when he made a major discovery. It was a piece of ocher, but marked with a cross pattern. Many now consider it the oldest work of art ever found. The first humans had managed, for the first time in prehistory, to store something outside their own brains, and in an artistic way, not just by engraving the rock (Fig. 1.7). They sent us a color message from 73,000 years ago! Other simpler motifs, also found in South Africa, were added to the discovery, with stone engravings and shells dating back to 100,000 years ago. And other pieces of jewelry were also found elsewhere that suggested that Neanderthals expressed themselves through art before *Homo sapiens* even reached Europe (Fig. 1.8).

To sum up this research on the above questions:

Fig. 1.8 Lebombo bone, from various angles. It is almost 44,000 years old. The 29 notches (one is missing, which is attributed to the fact that the end is broken) correspond to the days of the lunar month, easily associated also with the menstrual cycle. Reprinted with permission (PNAS) from: Francesco d'Errico, Lucinda Backwell, Paola Villa, Ilaria Degano, Jeannette J. Lucejko, Marion K. Bamford, Thomas F.G. Higham, Maria Perla Colombini, Peter P. Beaumont, Early Evidence of San material culture represented by Organic Artifacts from Border Cave, South Africa, 13,214–13,219, PNAS, August 14, 2012, vol. 109, no. 33, www.pnas.org/cgi/doi/10.1073/pnas. 1204213109Supporting Information Appendix, Fig. 8

1. The capabilities of hominids evolved much more *gradually* than previously thought, say in the 1970s, when it was believed that there was a sudden and very dramatic genetic change 50,000 years ago which resulted in humans being able to think and communicate.
2. Neanderthals and their common ancestors probably communicated orally with those of *Homo sapiens*, albeit in some rudimentary way. This would bring the origin of the first *oral expressions* to over two million years ago.
3. And, quite probably, the first documented mental capacity, acquired perhaps as early as eight hundred thousand years ago, was *geometry*. An incipient geometry, if you like, comprising simple stripes on stones and shells, but with very clear recurring patterns, found in various places, far removed from one another, over at least 73,000 years. And this was still several tens of millennia before graphic symbols were used to encode sounds, names, and languages.

1.2 The Oldest Proofs of Art and Mathematical Reasoning

But let us now focus specifically on the most remote tests of mathematical thought that have ever been found. This time, I mean, of an already more complex reasoning than that shown by the simple geometric patterns we saw before. The latest dates of the oldest bones yet found by archaeologists, featuring marks made on purpose, point to at least 43,000 (Lebombo), 30,000 (Wolf), and 20,000 (Ishango) years ago (Figs. 1.8, 1.9, 1.10 and 1.11).

They prove beyond any reasonable doubt the ability of our primitive ancestors to keep count of the menstrual cycle and of the phases of the moon, and later, of the beginning and subsequent development in their brains of certain mathematical capacities: first, of enumeration and grouping, and later of more complex calculations. As I will argue in various places in this book, we generally tend to underestimate the cognitive and manual abilities of our predecessors. Proofs of this fact are being reported, now and again, in highly prestigious journals, such as Nature and Science Advances. As already mentioned, in a letter to Nature on October 2018 [3], a cross-hatched pattern drawn with an ochre crayon on a ground silcrete flake was reported

Fig. 1.9 Of about the same epoch as the Lebombo bone: A 40,000-year-old bull painting, made with ochre, discovered in Lubang Jeriji Saléh cave, East Kalimantan, Borneo, Indonesia. At the moment of the discovery, it was the most ancient sample of figurative cave painting. Later, as referenced in the text, a 45,500-year-old representation of a Sulawesi warty pig has been found in Leang Tedongnge, Indonesia. And in an article published in Science, June 2018, a 64,800-year-old Neanderthal painting from La Pasiega cave, in Cantabria, Spain, was reported

Fig. 1.10 The Ishango bone on exhibition at the Royal Belgian Institute of Natural Sciences. Reprinted under license CC BY-SA 3.0

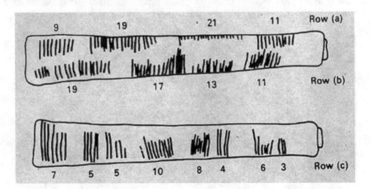

Fig. 1.11 Schematic representation of the Ishango bone. Contains grouped notches whose precise meaning is unknown. Reprinted with permission (Springer) from: Pejlare J., Bråting K. (2019) Writing the History of Mathematics: Interpretations of the Mathematics of the Past and Its Relation to the Mathematics of Today. In: Sriraman B. (eds) Handbook of the Mathematics of the Arts and Sciences. Springer, Cham. https://doi.org/10.1007/978-3-319-70658-0_63-1, Fig. 1

Fig. 1.12 Mesopotamian clay tablet with engraved Pythagorean triples, from ca 1800 BC. (Plimpton 322)

as the oldest indicator of modern cognition and behavior, recovered from approximately 73,000-year-old Middle Stone Age levels at Blombos Cave, South Africa. This is some 30,000 years older than previous findings, and no wonder that experts are still debating on how sound this conclusion is, and on whether the pattern can really be counted as a piece of art. But more recently, in January 2021, the uranium-series dating of two figurative cave paintings of Sulawesi warty pigs were reported in Science Advances [4]. These were found in Indonesia, one of them in Leang Tedongnge, with a minimum age of 45,500 years, and the second in Leang Balangajia, dated to at least 32,000 years ago. The authors of the find consider the animal painting from Leang Tedongnge to be the earliest known *representational* work of art in the world. I am sure that more, ever older findings will keep appearing, always pushing back this frontier.

1.3　First Scientific Concepts and Models of the Cosmos

Restricting ourselves to more specifically cosmological concepts, notions about them can be found in the oldest books ever written, either on clay tablets, or on parchment and papyrus.

In the West, when referring to really ancient texts, we tend to always think of some of those that later constituted the Bible, such as the Book of Job, whose first version is usually dated to around two thousand years

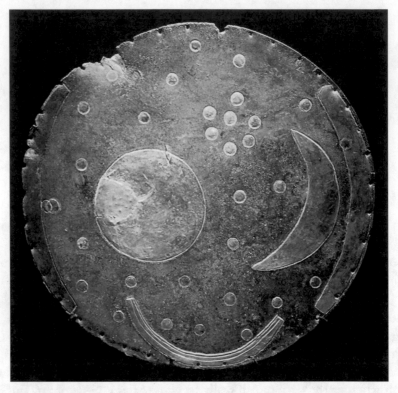

Fig. 1.13 Nebra sky disk 1600 BC, oldest concrete depiction of the cosmos yet known from anywhere in the world. In June 2013 it was included in the UNESCO Memory of the World Register and termed "one of the most important archaeological finds of the twentieth century". CC BY-SA 3.0

Fig. 1.14 Academy of Athens, under the watchful eye of Plato and Socrates. By ArmAg. Created: 10 September 2019. CC BY-SA 4.0

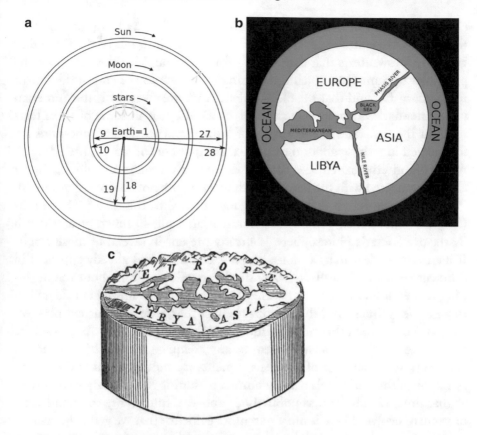

Fig. 1.15 Peter Apian's 1524 representation of the universe, heavily influenced by Aristotle's ideas. The terrestrial spheres of water and earth (shown in the form of continents and oceans) are at the center of the universe, immediately surrounded by the spheres of air, and then fire, where meteorites and comets were believed to originate. The surrounding celestial spheres from inner to outer are those of the Moon, Mercury, Venus, Sun, Mars, Jupiter, and Saturn, each indicated by the corresponding symbol. The eighth sphere is the firmament of fixed stars, which include the visible constellations. The precession of the equinoxes caused a gap between the visible and notional divisions of the zodiac, so medieval Christian astronomers created a ninth sphere, the Crystallinum which holds an unchanging version of the zodiac. The tenth sphere is that of the divine prime mover proposed by Aristotle (though each sphere would have had an unmoved mover). Above that, Christian theology placed the "Empire of God". File: Ptolemaicsystem-small.png. Created: 28 December 2005

BC, that is, at the time of the biblical patriarchs. This is half a millennium before the book of Moses, at the time when the Book of Genesis is usually located (although it should be noted that there is still no complete agreement on these dates). However, there are plenty of other writings, by Sumerians, Egyptians, and Akkadians, in particular, which according to reliable sources precede them by several hundred years, such as the texts of Abu Salabikh

(2600 BC), those of the Pyramids (2400 BC), the Enûma Eliš (1800 BC), or the famous Gilgamesh epic (1700 BC), to cite just four out of the fifty other currently known texts that date from before the Iron Age. Note also that the principles of Sumerian cuneiform writing date back to the end of the fourth millennium BC and that the oldest calendar we are aware of is the Sumerian lunar calendar, which dates back to ca 2700 BC. It is in some of these texts that, for the first time in the history of humanity, theories and questions are formulated in writing about the essential components of *the All*, including all material and ethereal entities that we can observe around us (albeit, almost always connected with the beyond, with what we are not able to see or touch).

Without going into details, I limit myself here to the theory of the four (five) elements which, although elaborated later and in much greater depth by the pre-Socratic philosophers, is already present in several of those works. It has become clear that, a thousand years earlier, it had already appeared in different places and cultures. Many fundamental questions about the beginning and end, and about the nature of the world were already considered in those ancient times. And this, despite the fact that, from our current perspective, we would think that those cultures could certainly not have possessed the knowledge required to answer them in any adequate way. But it was not in vain that, in the absence of good instruments for measurement and observation, one of the main tools namely human reasoning, was already well present at that time, having had, as pointed out above, a full twenty thousand years or more to evolve. There is more and more evidence that we generally tend to underestimate the knowledge and capabilities of those who came before us. I could give many examples, but for the sake of brevity, I shall not continue along this path (Figs. 1.12, 1.13 and 1.14).

We may just note that the pre-Socratic philosophers already had a good number of concepts as fundamental as those of substance, number, power, infinity, movement, being, atom, space, and time. The Greek *ta mathemata*, a plural noun used often by Plato, designated "what can be learned" and thus, at the same time, "what can be taught". And we could affirm (for this and several other important reasons) that actually math preceded philosophy itself as the first discipline from which all knowledge was derived.[2] The four *mathemata* were arithmetic, geometry, astronomy, and music [5] (which later evolved as disciplines of the Roman *trivium* and *quadrivium*). More to the point, it is known that Plato wrote at the entrance to his academy: *"Let no one ignorant of geometry enter here,"* while, a hundred years earlier, the Pythagorean School already had as its maxim that *"all things are numbers"*. As

---

[2] On this point, I have had several discussions, some of them endless, with fellow philosophers, which I will not reproduce here.

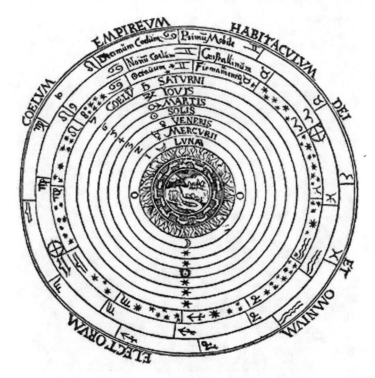

Fig. 1.16 **a** Map of Anaximander's universe (sixth century BCE). User: Bibi Saint-Pol—Own work (based on the text and GIF by Dirk L. Couprie for The Internet Encyclopedia of Philosophy: see http://www.iep.utm.edu/a/anaximan.htm#SH6h). **b** Possibly what the lost first map of the world by Anaximander looked like. User: Bibi Saint-Pol—Own work based on Anaximandermap.png. **c** Anaximander cylindrical Earth. *Source* Popular Science Monthly Volume 10

is well known, this historical epoch represents the triumph of knowledge, in all of its splendor.

But turning to cosmology, the first scientific model of the Universe, in the sense that it had far less mythological content than those that had preceded it, was built two centuries earlier. It was due to Anaximander (610–546 AC), born in Miletus, a disciple of Thales who continued his master's work. For the first time, his model dispensed with Atlas, who had hitherto always carried the enormous weight of the Earth on his back, preventing it from plunging into the depths of the abyss. In Anaximander's model, the Earth, a flattened cylinder of perfect proportions, was already floating freely in the ether (Fig. 1.16).

His model is particularly interesting. It contains an extraordinarily accurate description of the shapes, proportions, and distances of the heavenly bodies. In total agreement with the theory of the four elements, the Sun is located in

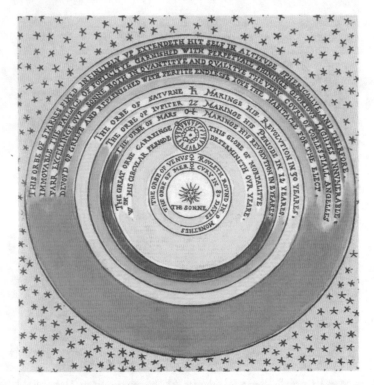

Fig. 1.17 An illustration of the Copernican universe from the book by Thomas Digges (1546?–1595). Created: sixteenth century date QS:P, + 1550-00-00T00:00:00Z/7

the furthermost circle, since it is fire, the largest fire; and fire always goes up. The circle of the Moon lies below, while the stars and planets are the smallest fires and rotate in inner circles within a cylinder of perfect proportions, like those of the entire model. It is a fascinating universe that dramatically highlights the extraordinary difficulties astronomers had, in those times, to assess the distances to the celestial bodies. (For more about this, the reader should consult the reference given above.) In fact, the representation by Thomas Digges of Copernicus' Universe, dating from the year 1576, was the first representation of the Universe in which the stars were already clearly arranged, all of them, beyond the circles of the planets (Figs. 1.15, 1.17 and 1.18).

Anticipating what will come later, it is worth mentioning that it was not until 1986 (just 35 years ago!) that the first three-dimensional map of the Universe was drawn. Actually, it was just a narrow slice of our neighborhood; but *in depth*, that is, giving the calculated distances to celestial objects (still with substantial errors, by the way). Over one thousand objects where plotted (Fig. 1.20). Until that very day, all the maps of the Universe (including

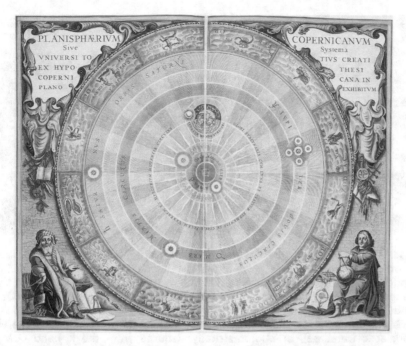

Fig. 1.18 Image of the new order of the cosmos proposed by Nicolas Copernicus, in which the Earth orbits around the Sun. Author: National Geographic Historia. CC BY-SA 4.0

the famous *APM Galaxy Survey*, with two million galaxies) were reduced to mere projections on the celestial sphere (Fig. 1.19). Projections are what Anaximander actually saw and what we all see at night, whether we use our eyes or a telescope, no matter how powerful: distances cannot be calibrated with the naked eye. All these considerations show, for all it may surprise the reader, that calculating distances in astronomy (not to mention cosmology) is extremely difficult; much more so than anyone would imagine, unless they happen to be specialists in this subject. The errors made throughout history in this calculation have been abundant and egregious, although henceforth not so much, of course, as those evidenced by Anaximander. Suffice it to add that, in 1929, in the calculation of his famous law, Hubble still made an error of close to 1000%, and note that this was just 90 years ago and that he was the best specialist of his generation in calculating distances.

It is commonly accepted that modern cosmology, as a science with capital letters, began to take shape at the dawn of the last century. And this happened in parallel with the advent of the aforementioned scientific revolutions, which provided—as I have already claimed—its theoretical foundation. The vision we now have of the global, or large-scale universe (what astronomers call the

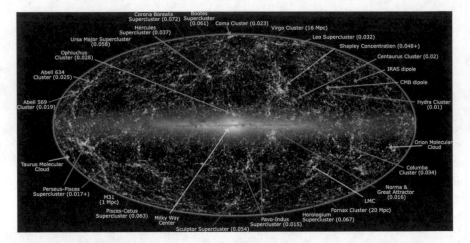

Fig. 1.19 Panoramic view of the entire near-infrared sky, revealing the distribution of galaxies beyond the Milky Way. The image is derived from the 2MASS Extended Source Catalog (XSC)—more than 1.5 million galaxies, and the Point Source Catalog (PSC)—nearly 0.5 billion Milky Way stars. The galaxies are color coded by redshift (numbers in brackets) obtained from the UGC, CfA, Tully NBGC, LCRS, 2dF, 6dFGS, and SDSS surveys (and from various observations compiled by the NASA Extragalactic Database), or photometrically deduced from the K band (2.2 μm). Blue/purple ones are the nearest sources (z < 0.01), green ones are at moderate distances (0.01 < z < 0.04), and red ones are the most distant sources that 2MASS resolves (0.04 < z < 0.1). The map is projected with an equal area Aitoff in the Galactic system (Milky Way at center). Date 2004. *Source* "Large Scale Structure in the Local Universe: The 2MASS Galaxy Catalog", Jarrett, T.H. 2004, PASA, 21, 396. Author IPAC/Caltech, by Thomas Jarrett

extragalactic universe) began to take shape from the 1910s to the 1930s. It should be added that, just a hundred years ago, all scholars were still absolutely convinced that the entire Universe was reduced to our galaxy, the Milky Way. Although many nebulae had already been detected before, no one had yet recognized them as objects located beyond our galaxy. In fact, the first nebulae were identified by Ptolemy, in his *Almagest*, in the year 150 of our era [6]. Later, Persian, Arabic, and Chinese astronomers recorded the existence of some others, over several centuries. Then, in scientific publications, Edmund Halley reported six in 1715 [7], Charles Messier cataloged 103 in 1781 [8], and between 1786 and 1802, William Herschel and his sister Caroline successively published three catalogs, containing a total of 2510 nebulae [9]. Of course, all were convinced that these were star clusters which simply could not be resolved with the telescopes of the time. It was Ernst Oepik [10] and Edwin Hubble [11] who, between 1922 and 1924, realized that some of the nebulae—like Andromeda and the Triangle—were definitely far beyond the Milky Way. In doing so, they suddenly changed the vision that

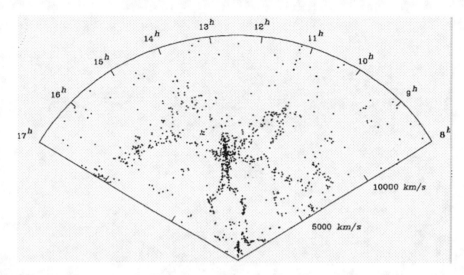

Fig. 1.20 Map of the observed velocity plotted versus right ascension in the declination wedge 260.5 ≤ δ ≤ 320.5. The 1061 objects plotted have mR ≤ 15.5 and V ≤ 15,000 km/s. From "A slice of the universe", de Lapparent, V., Geller, M. J., & Huchra, J. P. Astrophysical Journal, Part 2—Letters to the Editor (ISSN 0004-637X), vol. 302, March 1, 1986, p. L1–L5, Fig. 1a. http://adsabs.harvard.edu/full/1986ApJ...302L...1D. *Courtesy* V. de Lapparent and AAS

all astronomers had previously had of the Universe, opening up to human knowledge the much more complex and extended extragalactic Universe [12]. It is at that time that my brief story of cosmology will begin; and this coincides with the moment when cosmology finally could become a true scientific theory. It will thus be useful to devote the next chapter to explaining in detail what is meant by a scientific theory and what is not.

2

What Is a Scientific Theory?

Guided by the prestigious *Stanford Encyclopedia of Philosophy* [13], we learn that there are three different ways of looking at a scientific theory: the syntactic, the semantic, and the pragmatic. Wade Savage distinguishes these philosophical perspectives as follows [14]. The *syntactic* view that a scientific theory is an axiomatized collection of phrases or postulates has been challenged by the *semantic* view, according to which it is in fact a collection of non-linguistic models. Although this is about the meaning of the statements, not just their formal structure, it is still "linguistic" in the sense of linguistically describable. And both conceptions are challenged by the *pragmatic* perspective, which claims that a theory is an amorphous entity that, even though it consists of phrases and models, the examples, problems, standards, abilities, practices, and trends are equally important.

Thomas Mormann characterizes the syntactic and semantic visions in similar ways [15], and was one of the first to use the term "pragmatic vision" to capture the essence of the third perspective. The three approaches have been so named on the basis of a corresponding linguistics trichotomy that has its roots in the work of Charles Morris, a famous American philosopher and semiotician [16], and later in the work of Charles Peirce [17], who is sometimes known as "the father of pragmatism."

I consider it appropriate to make these reflections on the scientific method, a concept that is so important today, when there have been recurrent discussions at the highest level, some of them by physicists and mathematicians from various disciplines. There have even been discussions about "the end of

© The Author(s), under exclusive license to Springer Nature
Switzerland AG 2021
E. Elizalde, *The True Story of Modern Cosmology*,
https://doi.org/10.1007/978-3-030-80654-5_2

science," in part starting from the fact that the frequency of theoretical physicists awarded the Nobel Prize in Physics has been steadily declining since the golden age of the 1920s (John Horgan, National Geographic, 2014) [18]. In this context, it is worth mentioning various documents that are highly critical of the current evolution of scientific theories, such as those by Peter Woit [19] and John Horgan [20] , and the latter's blog, also brimming with criticism. Drawing on my personal experience, rather extreme and antagonistic examples come to mind (some of which I give below) of the two sides or components of the scientific method. In short, observation vs theory, or theory vs observation—since citing one before the other could already be interpreted as favoring one of the two, when it should not be so.

From a very different point of view, without going further into this profound debate, let us focus only on its most positive aspects, asking ourselves, to begin with: how have we come to such deep and elaborate definitions? The point is that this subject is not a simple one. I just mention it here for those who wish to look a little further into it. Let us first summarize a bit of history, and then briefly address the essence of the concept, taking care to introduce some clear examples that will serve as a firm guide to avoid getting lost among the definitions and concepts. In these examples, we will be able to make out the various features of the three perspectives of the more precise definition of a scientific theory given above (Fig. 2.1).

2.1   A Little History

The beginning of the history of the scientific method took place at the time when the two great cultures of the Ancient Greece and Persia met. This occurred precisely as their respective armies were engaging in battles that will always be remembered (Marathon, Thermopylae, Salamis, Platea). It is easy to forget that these two cultures had deep mutual respect for each other and often interchanged ideas and knowledge. It is well known that knowledge has no borders. But, while Babylonian, Indian, and Egyptian astronomers, as well as their physicists and mathematicians, formulated empirical ideas, it was the Greeks who first developed what we now know as the scientific method. However, we must recall that, initially, the philosophers of ancient Greece did not believe in empiricism, and they saw the act of measurement as the domain of artisans, even when it concerned geometry. Philosophers, including Plato, believed that all knowledge could be obtained through pure reasoning, and that nothing needed to be measured. This may sound strange to us now, but they had good reasons to think so, which it would take too

Fig. 2.1 Marble bust of Aristotle (384 BC–322 BC). Roman copy after a Greek bronze original by Lysippos, from 330 BC; the alabaster mantle is a modern addition. After Lysippos–Jastrow (2006). Public Domain

long to discuss at this point (I refer the reader once again to the Stanford Encyclopedia of Philosophy) (Fig. 2.2).

Measurement and observation, the foundations on which science is built today, were the contribution of Aristotle, rightly considered the father of science. He was the first philosopher to realize the importance of empirical measurement, and he established that knowledge could only be obtained from what is already known. He proposed the idea of induction as a tool for gaining knowledge, and understood that abstract thinking and reasoning must be based on findings in the real world. He applied his method to almost everything from poetry and politics to astronomy and natural history. His "proto-scientific method" consisted of making detailed observations in each case. Thus, to study the natural world, he scrutinized more than a thousand species and, in a treatise on politics, he studied the constitutions of the

Fig. 2.2   Portrait of Galileo Galilei (1564–1642) by Justus Sustermans, 1636

nearly one hundred and sixty Greek city-states. This was a gigantic under-taking, in stark contrast to Plato with his idea of a perfect republic, which was based solely on abstract theoretical notions of perfection, rather than on actual systems already up and running (Figs. 2.3 and 2.4).

The early stages of Islam constituted a golden age for the spread of knowl-edge, and the history of the scientific method should have enormous respect for some of the brilliant Muslim philosophers of Baghdad and Al-Andalus. On the one hand, they preserved the knowledge of the Greek philosophers, beginning with Aristotle; and on the other, they also added much of their own and became the catalysts for the formulation of a scientific method later recognized by modern scientists and philosophers.

Roger Bacon (1214–94) was one of the first European scholars to perfect the scientific method. He developed the idea of making observations, formu-lating a hypothesis, and then experimenting to demonstrate it. In addition, he meticulously documented his experiments so that other scientists could repeat them and thus verify their results. More than three centuries later, Francis Bacon (1561–1626) was one of the main engines of the development of the

Fig. 2.3 Portrait of Gottfried Wilhelm Leibniz (1646–1716) by Christoph Bernhard Francke (ca. 1695)

method. He reiterated the importance of induction as part of the scientific method, affirming that any new discovery should proceed through a process of observation, experimentation, analysis, and inductive reasoning, and then apply the findings to the universe in general. He also claimed that experimental evidence should be used to eliminate conflicting theories and get closer to the truth. The great philosopher and mathematician René Descartes (1596–1650), on the contrary, firmly defended the idea that the universe was like a giant machine. Consequently, if someone were able to understand the basic laws of the universe, he or she could then deduce how it would behave in any specific circumstance.

Fig. 2.4 Portrait of Isaac Newton (1642–1727). Godfrey Kneller—http://www.phys.uu.nl/~vgent/astrology/images/newton1689.jpg

2.2   Galileo, Newton, Leibniz

And so, we finally arrive at Galileo Galilei (1564–1642), remembered above all for his experiment in the Leaning Tower of Pisa, and who contributed decisively to the scientific method. The greatest physicists, like Albert Einstein, Stephen Hawking, and others have proclaimed him as the "father of modern science." There is no doubt that his methodology shaped physics and other fields of science based on mathematical theorems. Galileo taught us that science must stand on two fundamental pillars, namely, the experimental observation of nature and the scientific theory or fundamental law itself. If either fails, there is no science. No matter how beautiful a scientific theory, no matter how "natural" or "evident" it appears to our eyes in a thousand aspects, if it is not endorsed by the observation of nature, after very precise experimentation, it will be worthless (Fig. 2.5).

His methods, which originated the division between science and religion (until then always practically one and the same thing, also in the Islamic world) included a standardization of measures that allowed experimental results to be checked anywhere. His so-called principle of Galilean invariance, conveniently extended, became crucial to the formulation of Einstein's theories of relativity. My own graduate work and PhD thesis dealt largely with this

a
b

Fig. 2.5 Title pages of **a** Sidereus nuncius, 1610, with hand annotations, and **b** Discorsi e Dimostrazioni Matematiche Intorno a Due Nuove Scienze, 1638, by Galileo Galilei (1564–1642). Copyright: History of Science Collections, University of Oklahoma, http://www.libraries.ou.edu/info/index.asp?id=20. Pictures taken by the author with permission

question. In my work, I made a very thorough study of the extent to which Galileo's laws of invariance are contained in Einstein's theory of relativity; and also the precise relationship between one and the other theory, at the level of relations between their respective invariance groups [21]. Galileo affirmed that: *"Il libro della natura è scritto nella lingua della matematica"* (the book of nature is written in the language of mathematics). He was absolutely right, but he did not yet have the necessary mathematical tools, namely the calculus of infinitesimals, which was developed by Isaac Newton (1642–1727) and Wilhelm Leibniz (1646–1716) a few years later. The crucial importance that infinitesimal calculus (and indeed later developments in mathematics) has had in the evolution of all the sciences—not just physics (or natural philosophy, at that time)—is often overlooked or not properly recognized.

Galileo used a strongly inductive scientific method because he understood that no empirical evidence could perfectly match theoretical predictions. He rightly believed that it was impossible for an experimenter to account for all variables: no experiment could ever carry out perfect measurements, due to air resistance, friction, and inaccuracies in the temporary devices and methods employed. However, the repetition of an experiment by independent researchers could generate a set of evidence that would allow a correct extrapolation to be made of the general theory, of the fundamental law of nature. This period, spanning the sixteenth and seventeenth centuries, is often referred to as that of the scientific revolution. It reached its zenith with Isaac Newton, who was the first to understand that the scientific method needed both deduction and induction and that each could enrich the other (Fig. 2.6).

A major event that took place then was the foundation, in 1660, of the Royal Society. The purpose was to provide an expert group to advise and guide, but also to supervise the dissemination of information, and which moreover established a journal to aid in the development of the process. It ruled that experimental tests must always validate theoretical evidence, an idea that has become one of the foundations of modern science. The creation of expert groups and new journals also led to a true peer review system, a procedure already employed in the Islamic world.

2.3   Illustrative Examples

Having finished this historical part, it is now convenient to complete the clarification of the concept of a scientific theory with some specific examples that I consider appropriate to this purpose. But let us first summarize, as simply as possible, the very essence of the concept of a scientific theory. I will now extract what is common and absolutely key in all the formulations that we have mentioned above, starting in particular with Galileo. This should also guide us now and in the rest of this book. In practice, in any scientific theory, we distinguish two different parts (pillars or legs) on which it is supported, both being equally important and indispensable. On the one hand, results from the laboratory or direct observation of nature; on the other, the theoretical model or fundamental law these observations obey, providing the explanation of the results. No matter how elegant a model or a theory, if it does not correspond to observations of nature or laboratory results, this law or model will not have the slightest value as a scientific theory. On the other hand, a collection of experimental results or observations, no matter how accurate, careful, and exhaustive they are, will not have any value as a scientific theory if we are not able to determine the law or model, of more or less broad scope, which they obey.

Drawing on my personal experience, two extreme, antagonistic examples come to my mind that are located in the antipodes of the two components of the scientific method we are talking about: observation and theory, or theory and observation, in any order. In my various stays at the II. Institut für Theoretische Physik at the University of Hamburg, located in the grounds of

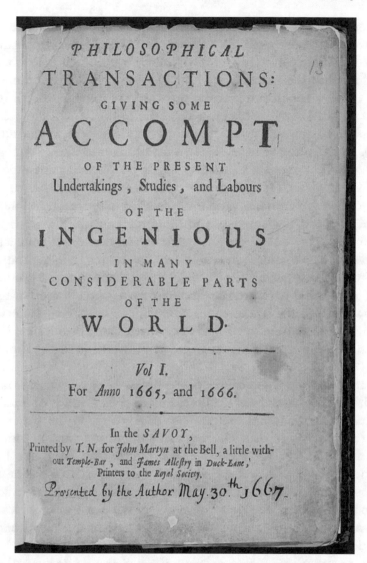

13

PHILOSOPHICAL
TRANSACTIONS:
GIVING SOME
ACCOMPT
OF THE PRESENT
Undertakings, Studies, and Labours
OF THE
INGENIOUS
IN MANY
CONSIDERABLE PARTS
OF THE
WORLD.

Vol I.
For *Anno* 1665, and 1666.

In the *SAVOY*,
Printed by *T. N.* for *John Martyn* at the Bell, a little with-
out *Temple-Bar*, and *James Allestry* in *Duck-Lane*,'
Printers to the *Royal Society.*

Presented by the Author May. 30.ᵗʰ ⎰667.

Fig. 2.6 Title page to volume 1 of Philosophical Transactions. Created: 28 November 1666. Henry Oldenburg - Philosophical Transactions. CC BY 4.0

the Deutsches Elektronen-Synchrotron (DESY), in the district of Bahrenfeld of the Hanseatic City, I spoke several times with Prof. Samuel C. C. Ting, a famous American physicist at the Massachusetts Institute of Technology (MIT), in the USA, recipient of the 1976 Nobel Prize in Physics. Professor Ting is a strong advocate of experimentation in particle physics and claims that theorists serve little purpose: it is the results of scrutinizing nature that count and it is upon them that physics must be built and make progress. At the opposite extreme in the 'confrontation' that I intend to highlight here, is Edward Witten, a professor of physics and mathematics at the Institute for Advanced Studies at Princeton, in the USA. He is one of the main champions today of the theory *par excellence*, within at least the domains of fundamental physics: string theory (or superstring theory, as we come to call it). It would be difficult for Professor Witten to receive the Nobel Prize in Physics, but he does possess an equivalent award in mathematics, the Fields Medal, attributed in 1990. At this other extreme, we find the proclaimed *Theory of Everything* or *M Theory* (for *membrane, matrix, magic* or, according to some, *mysterious*). Witten is one of the main actors in this field, which has turned out to be enormously fruitful in various domains of mathematics. Moreover, its potential to eventually solve some burning questions in physics is undeniable, but until now no one has been able to establish direct connections of any relevance to laboratory physics. In fact, interesting unifying connections among totally different fields of physics which seemed to have nothing to do with each other have been highlighted through the use of string theories. However, it is the lack of connection with experimental results that has led many physicists from other fields to criticize those theories (and the fact that such large amounts of money have been invested in them), for multiple reasons (Figs. 2.7 and 2.8).

Let me mention another couple of examples. As of now, numerous astronomical missions have already collected a fabulous amount of data on the existence, distribution, and evolution of dark matter in the Universe. The quantity and quality of this data has been accumulating and improving day by day, for several decades. However, no one has yet been able to figure out a theoretical model that can irrefutably explain all these data, that is, with a precision of several sigmas. Therefore, we do not yet have a scientific theory on dark matter. We are missing one of the two legs on which such a theory should stand: the theoretical model or fundamental law, that is, the explanation of the nature and behavior of dark matter.

Going to the other extreme, we have also had around, for several decades already, a spectacular theory known as supersymmetry. As a theoretical model, it could not be brighter, more attractive, and well elaborated. Many physicists (myself among them) think that this theory must necessarily be correct, that it cannot be false. But behold, to date there are still no experimental results to confirm it, however many attempts have been made to find them with various experiments at the Large Hadron Collider (LHC) at CERN (in Switzerland) and in other high-energy laboratories around the world. And thus, supersymmetry is not currently a scientific theory. It lacks the other pillar: the observational one.

Fig. 2.7 A photo of Nobel laureate Dr. Samuel C. C. Ting taken on October 19, 2010 after a presentation he made on the Alpha Magnetic Spectrometer to employees at the Kennedy Space Center in Florida. *Source* Toastforbrekkie—Own work, 2010. CC BY-SA 3.0

Fig. 2.8 Edward Witten giving a speech at the Chalmers Tekniska Högskola, Göteborg, Sweden. *Source* Ojan–Own work, 2008

The Higgs model, according to which the constitutent mass of elementary particles is generated by spontaneous breaking of a symmetry, and which predicted the existence of the Higgs boson, is another wonderful example of a treasured theory that did not become scientific until experimental evidence of the Higgs boson appeared in 2012 at the LHC. This was almost fifty years after Peter Higgs and others formulated it, in 1964. Other examples, equally magnificent, are those of Paul Dirac and Wolfgang Pauli, who predicted, respectively, the necessary existence of positrons (and with these, antimatter) and neutrinos (which we now know are ubiquitous in the Universe). They achieved this on the basis of their purely theoretical models, which these two great scientists considered to be necessarily real, that is, true descriptions of the world in which we live. Precise experiments later proved they were both right.

To end this chapter, I will give yet another example. A very important one that is directly related to cosmology and in particular its conversion into a true science at the beginning of the twentieth century. Later I will elaborate on this point in much more detail, but I would like to present the conceptual idea here. It has to do with the discovery of the expansion of the universe. This is today a well settled fact in the framework of a scientific theory (the general theory of relativity), but regarding historical accounts of this discovery there is a huge (and sometimes perfectly incomprehensible!) confusion in the literature. The vast majority of books, encyclopedias, articles, and lecturers galore claim that "it was Edwin Hubble who discovered that the universe is expanding." But this is not true, for several reasons. The incontestable truth, discovered by science historians after much painstaking research, is that Hubble never believed in the expansion of our Universe, and that he even fought against this idea, despite the fact that he contributed so significantly to the discovery. You may now ask me, how is that possible? Did he discover the expansion of the universe or not?

The puzzle is not so hard to understand, if we stick to the definition of a scientific theory. Remember that:

Scientific Theory = Observation + Law or Theoretical Model (explanation).

Hubble only contributed to the first of the two legs: the *empirical* or *observational* part. He constructed a detailed table of distances to the spiral nebulae and, by comparing it with the table of radial recession speeds of these nebulae (work by Vesto Slipher), he found his famous law. This indicated that the nebulae were moving away from us at a speed proportional to their distance: the further they are, the faster they move away from us. This meant, without a doubt, that *the size of the visible universe was increasing*: the universe was getting bigger and bigger. This conclusion was crystal clear, as a purely empirical result. All astronomers quickly admitted it and none ever raised the least objection (well, this is simplifying too much, I will elaborate a lot on this point later). Hubble suddenly became very famous. But what I have said has *nothing* to do (at the outset) with saying that the universe expands, that is, that the fabric of space-time stretches, dragging these nebulae with it. And it is now well known that Hubble *never* believed that things happened this way. He always looked for a *completely different explanation* for the recession of the nebulae at those tremendous speeds. Neither he, in his entire life, nor Einstein, for ten long years (as we will see in detail later) were able to accept this concept of "universal expansion," which falls to the other side of the equation above. For it is precisely the fundamental physical law that is necessary to *explain* the recession of the nebulae in a natural and comprehensible manner, that is, why this recession happens exactly the same way in all directions, as if celestial bodies had all agreed to fly away from each other in the same way? The universal expansion is an extremely difficult concept to understand and accept. The reader should ponder on why Einstein, a true genius, the very creator of the general theory of relativity, insisted on dismissing this possibility so strongly for a whole decade!

In short, the scientific method requires a theory—a simple, beautiful, and general one, unique, if possible, within the scheme in question—and its experimental verification in the laboratory, with the greatest possible precision and by various groups, totally independently of each other.

As an example, the theory of relativity has more than satisfied all these criteria in a brilliant, never equaled way. Einstein started from pure logic, from unprovable but abundantly clear basic assumptions which he took as axioms, such as the equality of inertial mass and gravitational mass, and the constancy of the speed of light in the vacuum—verified in an experiment carried out by Albert Michelson in 1881 and then more precisely by Michelson and Edward Morley in 1887—then using a principle of relativity and building upon Bernhard Riemann's geometry, Einstein put together an extremely beautiful theory that has been proven true in all the experiments carried out to date, in our Solar System, in our galaxy, and in other, far more distant regions of the Universe.

All modern cosmology is based on Einstein's field equations, which he formulated in 1915, corresponding to the theory now known to physicists as General Relativity. This has led many authors to locate, with precision, the origin of modern cosmology (at least, of theoretical cosmology) in February 1917, when Einstein first used his equations to build a model that could describe our Universe. At that time, the Universe was believed to be static, without beginning or end, and it was rather small because everyone was convinced that all celestial objects, i.e., everything that could be seen in the night sky, was inside the Milky Way. However, such a static universe was not a solution to his field equations, so Einstein introduced an additional term: the now famous cosmological constant (we will come back to it later). With the appropriate sign, it provided a kind of repulsive force that counteracted the pull of gravity which would otherwise have caused his static universe to collapse, balancing it exactly. But now we have already gone too far in that direction, so I shall end these theoretical considerations here, only to continue later along this path.

3

The First Cosmological Revolution of the Twentieth Century

Our mission in this chapter already honors the title of the book. As good explorers, we shall now begin our search for the origins of cosmology as a modern science. Having learned the lesson from the preceding chapter, we must determine, clearly and precisely, the moment when scientists first found themselves in a position to study the Universe in its entirety, as they would any other physical system.

On a very large scale, that is, on a cosmic scale, celestial objects are points roaming through space. A glance at the first in-depth map of the Universe (Fig. 1.20) is enough to realize this fact. Depending on their distance from Earth, the points will be stars, galaxies containing hundreds of millions (or billions) of stars, or galactic clusters containing perhaps hundreds of billions of stars. On a cosmic scale, the Universe is a collection of points and, like any physical system, it is well specified if we know their masses, positions, and velocities. They can thus be represented in the so-called phase space. In contrast to a laboratory system, we always observe the universe from the same place, so the position of celestial objects will always be naturally given by the (radial) distance to the object and two angular coordinates of the radial line, or equivalently, of its projection on the celestial sphere. Likewise, the velocity vector of each object will have a radial component, zooming out or approaching us, and another component perpendicular to the radial direction, which determines its displacement on the celestial sphere (Fig. 3.1).

© The Author(s), under exclusive license to Springer Nature
Switzerland AG 2021
E. Elizalde, *The True Story of Modern Cosmology*,
https://doi.org/10.1007/978-3-030-80654-5_3

Fig. 3.1 Henrietta Leavitt (1868–1921)

Let us now repeat the initial question, our goal: when did cosmology first become a modern science? The answer is now clear: when tools first became available that allowed (men and women)[1] researchers to calculate both:

(i) **positions** and
(ii) **velocities**

of the celestial bodies on a large scale. We will see next at what precise moment these two fundamental tools became available to students of the cosmos, and this will give us the exact date of the beginnings of cosmology as a true science. It's that easy. We will also realize that the knowledge obtained through the acquisition of these tools gave rise to a true revolution in our understanding of the cosmos, which shattered the previously held model of the Universe. I should emphasize, although this is obvious, that I am talking here about very distant objects. When it comes to studying celestial objects close to Earth, such as the Moon, the planets, the Sun, or even nearby stars, the enormous simplification I am making will no be longer useful, and nor would it even hold up. We then come under the auspices of astrophysics,

---

[1] To simplify, I will not always use the double gender, which I will consider implicit. My emphasis on the discoveries of both female and male astronomers and male and female cosmologists should sufficiently demonstrate my neutrality on this issue.

an extraordinary science whose objectives overlap with those of cosmology in intermediate domains of the cosmos, but which I will not discuss in my analysis, limited here to gigantic scales.

3.1 What Was the Universe Like a Hundred Years Ago?

With our objective clearly established and following the historical line of the opening chapter, the question is now: what was the Universe like a hundred years ago? In other words, how far had the knowledge of astronomers, philosophers, and theorists gone in representing the cosmos by then? There was no doubt at the time that enormous progress had been made in the knowledge accumulated by the various sciences, to the point that, towards the end of the nineteenth century, some famous physicists had proclaimed that all the important laws and fundamental principles had probably already been discovered, and that the only thing left to do was "adjust the results to the sixth decimal number" (Albert Michelson). Others were aware that there were still a few adjustments to make on the fringes, "a pair of little clouds" that resisted attempts to remove them (Lord Kelvin). What is certain is that no one thought they could be on the very doorstep of the greatest revolutions that science would ever see, those originated by Planck and others as early as 1900.

In the case of the Universe, the model accepted by everybody—obviously much more advanced than the one discussed in Chap. 1—had the following properties:

1. **The Universe was eternal, without beginning or end**. There was no scientific evidence that it could have originated at any point in the past, nor was there reason to think that it would have an end at any point in the future. Science and religion had been completely separated by then, as mentioned before. And, in the domains of pure reason, the biblical genesis and apocalyptic tales had no place in scientific explanations (Fig. 3.2).
2. **The Universe was static, or stationary**. This does not mean that celestial objects did not move, but that, on a large scale, as a point dynamical system, celestial bodies would behave according to the well established laws of mechanics and thermodynamics. These state that, under very general conditions, a physical system evolves over time into a stationary state. And, according to the preceding point, the Universe had had infinite time to evolve to it. Its state could not be other than stationary. In

Fig. 3.2 Harvard Computers at work, circa 1890, including Henrietta Swan Leavitt seated, third from left, with magnifying glass (1868–1921), Annie Jump Cannon (1863–1941), Williamina Fleming standing, at center (1857–1911), and Antonia Maury (1866–1952). File: Astronomer Edward Charles Pickering's Harvard computers.jpg. Harvard College Observatory. Created: circa 1890

other words, one would always and forever see the Universe in the same way.

3. **The Universe was reduced to the Milky Way**. It was therefore very small (insignificantly small, in current terms). I have already discussed the perennial problem of the precise calculation of distances which astronomers faced then, and which is still the main issue in cosmology today. Galaxies, which had already been observed by their thousands and which were called nebulae, were all believed to lie within the Milky Way, the only galaxy recognized as such at the time.

This model of the universe was about to change thoroughly and radically, in all its conceptions. As I always put it in my papers and talks, what I have baptized as the first revolution in cosmology of the 20th century was about to take place. And it coincided exactly with its conversion into a modern science.

3.2   Henrietta Leavitt

By now the reader should have grasped how extremely difficult it is to calculate distances in astronomy. The order of magnitude of the errors made in this regard during past epochs is simply colossal. Furthermore, this is precisely why the first hero in our story—in fact, a she-hero, a heroine—must undoubtedly be Henrietta Leavitt, such is the importance of the extraordinary discovery she made in 1912, after several years of collecting thousands of data, in particular from the Magellanic Clouds, satellite galaxies of our Milky Way. This was the discovery of the period-luminosity relationship of Cepheid variable stars: a linear dependence of the real luminosity versus the logarithm of the period of variability of the luminosity of the star [22] (Fig. 3.3).

Henrietta Leavitt was born in Lancaster, Massachusetts. Her mother was a descendant of a Puritan English tailor, who settled in the Massachusetts Bay Colony in the early seventeenth century (in the earliest census records, the

Fig. 3.3 Photograph of the Harvard Computers (unflatteringly known as "Pickering's harem"), a group of women who worked under Edward Charles Pickering at the Harvard College Observatory. The photograph was taken on 13 May 1913 in front of Building C, which was then the newest building at the Observatory. The photo was discovered in an album which had once belonged to Annie Jump Cannon. Image courtesy of the Harvard-Smithsonian Center for Astrophysics. Back row (L to R): Margaret Harwood (far left), Mollie O'Reilly, Edward C. Pickering, Edith Gill, Annie Jump Cannon, Evelyn Leland (behind Cannon), Florence Cushman, Marion Whyte (behind Cushman), Grace Brooks. Front row: Arville Walker, unknown (possibly Johanna Mackie), Alta Carpenter, Mabel Gill, Ida Woods. Created: 13 May 1913

surname was spelled "Levett") [23]. Henrietta was always very religious. She studied at Oberlin College, which later became Harvard's Radcliffe College, earning her bachelor's degree in 1892. In her fourth year of college, she enrolled in an astronomy course, and then began working as one of the "women computers" at the Harvard College Observatory, hired by its director Edward Pickering. Her mission was to measure and catalog the brightness of the stars in the observatory's precious collection of photographic plates. Women were not yet allowed to operate telescopes. In this sense, she was just one among the large group of women that appear in various photographs of the time, clustered around the director. They have received various names, such as the "Harvard computers" or the more festive "Pickering's harem." [24]. In 1898 Leavitt became a member of the Harvard staff as the first "curator of astronomical photographs." After making two trips to Europe, she contracted a disease that made her gradually lose her hearing. She returned to Harvard in 1903 and is known to have earned ten and a half dollars a week at the time. She was reportedly "hardworking, serious, little given to frivolous activities and selflessly devoted to her family, her church and her career." Pickering assigned Leavitt to study the variable stars of the Magellanic Clouds, recorded on photographic plates taken at the Harvard Observatory in Arequipa, Peru. She identified 1777 variable stars and, in 1908, published these results in the Annals of the Harvard College Astronomical Observatory. She carefully noted that the brightest variables had the longest period of variability.

But her decisive work, central to the history of astronomy, came in 1912: Leavitt carefully examined the relationship between periods and brightness in a sample of twenty-five of the Cepheid variables from the Small Magellanic Cloud. The resulting document was communicated and signed by Edward Pickering, but in its first paragraph it is already indicated that it was in fact "prepared by Miss Leavitt." She plotted a graph of magnitude versus the logarithm of the period of variability and concluded that:

> A straight line can easily be drawn between each of the two series of points corresponding to the maximums and minimums, showing that there is a simple relationship between the brightness of the Cepheid variables and their periods.[15]

But, why was this discovery so fundamental? It was neither more nor less than one of the two extremely powerful tools that astronomy was looking for. By determining the period of variability of a Cepheid star, this relationship, now called "Leavitt's law," allows us to calculate its intrinsic luminosity. And, from the intrinsic luminosity and the luminosity we detect here on Earth, we can use Gauss' law to calculate the radial distance to the star. The cleverest reader will have observed right away that up to this point it was just a law of proportionality, that is, it still had to be calibrated to determine the absolute distance. In other words, it was still necessary to find the scale factor; and it turns out that the distances to the Magellanic clouds were still unknown at the time. Leavitt expressed the hope that the parallax could be measured to some of the Cepheids (a method used to calculate distances to stars that are not too far from Earth, which I will describe below). This soon happened, thus allowing Leavitt to calibrate the luminosity scale and finally obtain her law as such, in its final form (Fig. 3.4).

The measurement of distances by parallax is an application of the triangulation principle, which states that, in any triangle, all sides and angles can be found if, in addition to two of the angles, the length of at least one side is known. Measuring a baseline gives us the scale of the entire triangulation network. In stellar parallax measurements, the triangle is extremely long and narrow. One measures the shortest side, which results from the motion of the observer and which in our case is the diameter of the Earth's orbit around the Sun, and as the upper angle is very small (for stars it is always less than one arc second), the other two are almost 90 degrees, whence the lengths of the two long sides of the triangle can be treated with very little error as roughly the same. Under these conditions, the distance to an object (in parsecs) is the reciprocal of the parallax value (measured in arc seconds): d *(parsec)* $= 1/p$ *(arcsec)*. Thus, for example, the distance to Proxima Centauri is $1/0.7687 = 1.3009$ parsec ($= 4.243$ light-year).

Fig. 3.4 Henrietta Leavitt (1868–1921). American Institute of Physics, Emilio Segrè Visual Archives. Author: Margaret Harwood

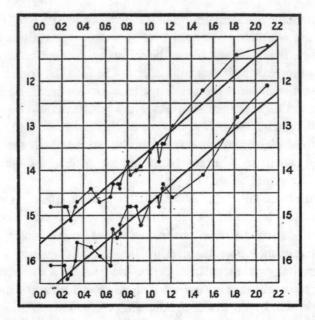

Fig. 3.5 Plot from a paper prepared by Leavitt in 1912. The horizontal axis is the logarithm of the period of the corresponding Cepheid, and the vertical axis is its magnitude. The lines drawn connect points corresponding to the stars' minimum and maximum brightness, respectively. From Leavitt, Henrietta S; Pickering, Edward C—"Periods of 25 Variable Stars in the Small Magellanic Cloud", Harvard College Observatory Circular, vol. 173, Fig. 2

For its simplicity and practical importance, Leavitt's discovery is one of the most important in the history of astronomy [25]. In the words of scientific writer Jeremy Bernstein: *"Variable stars had been of interest for years, but when he was studying those plates, I doubt Pickering thought he would make a significant discovery, one that would eventually change astronomy."* The period-luminosity law for Cepheids, aka Leavitt's Law, made these stars the first "standard candles" in astronomy, allowing scientists to calculate the distances to remote galaxies, where observations of stellar parallax are no longer useful for this purpose.[2] A year after Leavitt reported her results, Ejnar Hertzsprung determined the distance to various Cepheids in the Milky Way, and with this calibration of the scale, the distance to any Cepheid could thereafter be accurately determined (Fig. 3.5).

Cepheids were soon detected in other galaxies, such as Andromeda (Hubble, in 1922–23), and these provided an important part of the evidence

---

[2] The angle is practically zero, within the margin of error, and the distance to the star is practically infinite. It is impossible to specify the distance with any accuracy.

that spiral nebulae were independent galaxies located far from our own Milky Way. Leavitt's discovery would forever change our image of the Universe. As we will see, in the "Great Debate" of 1920, of which I will speak later, it induced Harlow Shapley to move the Sun away from the center of the galaxy, where it had always been supposed to sit, to a position further out, and it also induced Edwin Hubble to remove our whole galaxy from the center of the universe. And this, before Einstein's formulation of the general theory of relativity, which finally put everything in its rightful place. I will elaborate on these points in much more detail later on. Leavitt's discovery opened the way for the construction of modern astronomy, and for the study of the structure and scale of the Universe [26]. The great achievements that came thereafter were only possible thanks to Leavitt's law.

In 1921, when the renowned astronomer Harlow Shapley took over as director of the Harvard Observatory, Leavitt was appointed head of stellar photometry. Later that year, she succumbed to cancer and was buried on her family's plot in the cemetery in Cambridge, MA. She was just 53 years old. Hubble had said of her that she deserved the Nobel Prize for her work; and Gösta Mittag-Leffler, a member of the Swedish Academy of Sciences, and by then a very influential person in awarding the prizes, was in fact going to nominate her in 1924, when he discovered that Leavitt had died of cancer three years before.

Before ending this section, it would be useful to answer a question that more than one reader will already have asked: What is the physical explanation of Leavitt's law? Why on earth does there have to be a relationship between the luminosity of a variable star and its period of variability? The mechanism behind changes in the luminosity of many types of pulsating variable stars is called *kappa opacity* [27] (historically it was known by the more suggestive name of the "Eddington valve", but this term has become somewhat obsolete). Here, the Greek letter *kappa* (κ) is used to indicate the radiative opacity at any particular depth in the stellar atmosphere. In a normal star, an increase in the compression of the atmosphere causes an increase in its temperature and density. This in turn produces a decrease in the opacity of the atmosphere, which allows the accumulated thermal energy to escape more quickly. The result is a balanced condition, a balance between temperature and pressure. Instead, the non-adiabatic stellar pulsation resulting from the κ mechanism occurs in regions where hydrogen and helium are fully or partially ionized. An example of such a zone is found in the variable stars *RR Lyrae*, in which a second partial ionization of helium takes place, while hydrogen ionization is probably the cause of the pulsation activity in the variables *Mira* and *ZZ Ceti*. To oversimplify, the valve effect occurs as follows. Helium can be partially ionized (when it has lost a single electron) or totally (when it has lost both), which occurs in the layers of the atmosphere that are, respectively, furthest and closest to the surface of the star. When the star's atmosphere is compressed, much of the energy is expended in doubly ionizing helium, which transforms into He^{++}, but it happens that it then becomes opaquer, so that the radiation of the star, unable to escape, pushes it to the upper layers, until it cools down and becomes partially ionized He^+. But this helium is more transparent to radiation, so it finally lets light through and the luminosity peak of the variable star occurs. When the radiation has escaped and there is no longer enough to keep the He^+ layers away, these (being heavier) fall back onto the surface, heat up and become He^{++} again, more opaque and light, accumulating the radiation, which causes it to rise once more until it cools down and becomes He^+; which allows the radiation to escape again. It is an apparently simple mechanism, although it is not without problems which I cannot detail here (some concern the very principles of thermodynamics). In conclusion, suffice it to point out that various pulsation mechanisms have been described and that this continues to be an interesting topic of study in astrophysics [28].

In short, Leavitt's result provided an extremely powerful method for calculating distances. In fact, it was the main tool used by Hubble in the following years, and later on by several generations of astronomers, with remarkable success, until the arrival of improved techniques (Fig. 5.5), culminating in *SNIa* as standard candles. It was the latter that led to the discovery of an acceleration in the expansion of the universe (NP in Physics 2011). However, this is already the second cosmological revolution of the past century, which we will consider in another chapter.

3.3   Vesto Slipher

My second great hero in this story from a century ago is none other than astronomer Vesto Slipher. Starting in exactly the same year of 1912 in which Leavitt's fundamental work had appeared, Slipher began to develop a project, which would become transcendental, obtaining for the first time the radial velocity of a spiral nebula, Andromeda. The idea was to use the optical Doppler effect, determining as accurately as possible the changes in the spectral lines (Doppler shifts), either towards the blue or towards the red. He used the 24-inch telescope at the Lowell Observatory in Arizona. In some of my works [29], I have already stressed with great emphasis the enormous importance of Slipher's work, and it will be made clear again in what follows. But,

Fig. 3.6 V.M. Slipher, astronomer at Lowell Observatory from 1901 to 1954. Unknown author, Lowell Observatory. Created: 1 January 1909. CC BY-SA 4.0

when a few months ago I was asked by the Royal Spanish Physics Society, to write a comment for the society journal about the recent NP awardee James Peebles, and I had the magnificent opportunity to delve into his website from Princeton University, I had no idea what a huge surprise was awaiting me. I discovered the extent to which Peebles shared my opinion. It was a truly unexpected find. I realized that half of his rather succinct website[3] was devoted to what follows (Fig. 3.6).

"*Regarding the hypotheses of dark matter and quintessence, I draw attention to the verse.*[4]

So now, we are in Boston,
The home of the bean and the cod,
Where the Lowells talk only to Cabots,
And the Cabots talk only to God.

---

[3] That changed completely, however, two weeks after he was awarded the Nobel Prize.

[4] This is a popular poem, adapted from a famous toast someone made at a Red Cross meeting in Boston in 1910. It refers to two of the most influential Boston families.

One might be inclined to compare matter families that interact only with gravity with the Cabot family. But Percival, of the Lowell family, used his fortune to establish the Lowell Observatory and took the Slipher brothers there; their notable contributions include the discovery of cosmological redshift." (Figs. 3.7, 3.8 and 3.9).

The Lowell Observatory was undoubtedly key to the earliest origins of the cosmological revolution. No matter how much Edwin Hubble persisted in affirming throughout his life that it was the Mount Wilson Observatory alone that really got things going. And Hubble forgot to admit—and never did admit until the very year of his death—that it was Vesto Slipher who had opened his eyes to the conclusion that the Universe could not be static. Furthermore, the first step Slipher had taken was in fact the most important, making later progress in that direction, already open, relatively simple. Aside from Hubble's character, which was in many respects Slipher's antithesis, and from Slipher's extreme shyness in publishing his results or even presenting them at scientific conferences, there are several reasons why many of them which are now valued as really important went unnoticed at this time and even for many years after his death. The founder and first director of the observatory himself, the millionaire astronomer Percival Lowell, with his controversial ideas—which he did not hesitate to publicize widely—about the

Fig. 3.7 Peebles' work has deepened our understanding of the universe's structure and history. James Peebles official Nobel portrait. ID: 100,425. © Nobel Media. Photo: A. Mahmoud. With permission

Fig. 3.8 Clark Dome at Lowell Observatory. This is an image of a building, listed on the National Register of Historic Places in the United States of America. Its reference number is 66000172. CC BY-SA 3.0

Fig. 3.9 Mount Wilson Observatory, Angeles National Forest, California, US. Author: Craig Baker. CC BY-SA 4.0

possibility of finding life on Mars (that is, the builders of the famous channels that he himself had observed), also contributed decisively to the fact that the results obtained in his observatory were not properly appreciated by the scientists of the day. All this is collected in a very precise way in the magnificent biography of Lowell written by William Hoyt [30] (Figs. 3.10 and 3.11).

Vesto Melvin Slipher (whom everyone usually referred to as "VM") was born in 1875 in Mulberry, Indiana. Son of farmers, he studied at the university of his native state, where he graduated in mechanics and astronomy in 1901, and successively obtained a master's degree, in 1903, and a doctorate in 1909. In 1901, recently graduated, one of his teachers recommended him to Percival Lowell, who agreed, albeit with little enthusiasm, to hire him as his assistant for a limited time. His duties included taking care of Lowell's garden during his frequent trips to Boston, and also caring for Venus, the observatory cow, and her calves. Little did Slipher imagine then that he would spend fifty-three years of his life working in that place. When he married in 1904, he took his wife to Flagstaff. They lived in a house near the observatory. There, they had two children. Over time, he became increasingly involved in community life, in particular in the creation of the Flagstaff High School. Slipher was promoted to assistant director of the observatory in 1915. Upon Lowell's death in 1916, he was appointed acting director; and finally, director from 1926 until his retirement in 1954, at the age of 79. His younger brother

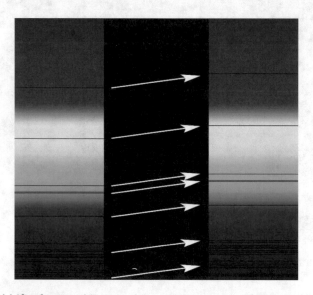

Fig. 3.10 Redshift of spectral lines in the optical spectrum of a supercluster of distant galaxies (right), as compared to that of the Sun (left). Georg Wiora (Dr. Schorsch) created this image from the original JPG. CC BY-SA 3.0

Fig. 3.11 Percival Lowell observing Venus from the Lowell Observatory in 1914. Reconstructed from several online sources by Joe Haythornthwaite. Unknown author. Public Domain

Earl Slipher (mentioned by Peebles, too) was also an astronomer. He joined him at the observatory and later served for a time as its director. He was also mayor of the city of Flagstaff and a member of the Arizona state legislature. Both brothers now name a crater on Mars.

The first commission Slipher received from Lowell was to assemble the spectrograph that Lowell had just purchased for the observatory. It made it possible to study the spectral lines of light coming from stars and other celestial objects. The Doppler effect had been known since the mid-nineteenth century. When a star moves away from Earth, its spectrum shifts toward the red (this is what is known as *redshift*), whereas if it approaches us, its spectrum presents a *blueshift*. For those who do not know, the spectral lines are the traces left by the different chemical elements in the light coming from celestial objects, and which are collected in the form of dark lines at certain wavelengths, typical of each element. If a light source approaches or moves away from us, the wavelength corresponding to this recorded footprint moves towards shorter wavelengths (which are bluer), or towards longer wavelengths (which are redder), respectively. This is completely analogous to the acoustic Doppler effect, which occurs when a noisy object (typically a motorcycle) first approaches and then moves away from us. As the motorcycle approaches, the frequency of the sound increases more and more (or equivalently, the wavelength decreases), then as the motorcycle moves away, the frequency abruptly decreases (the wavelength increases). As in the case of the motorbike, the optical Doppler effect tells us if a luminous celestial object is moving away or approaching, and the extent of the shift can tell us the relative speed. In fact, measuring this effect with precision is not an easy task, given the extreme faintness of the light emitted by very distant objects—for which the spectral shift should in principle be greater—and this is the first major obstacle. Slipher's observations showed that the greater the radial speed of the star, the greater the magnitude of its shift towards the relevant color, in perfect agreement with the formula for the optical Doppler effect. This point may seem quite obvious as explained here, but many cosmologists neither understood nor appreciated it for many years (Figs. 3.12 and 3.13).

Slipher's early research aimed to find out the rotational speed of certain spiral galaxies, including Andromeda, and discovered—using complex photographic techniques that required exposure times as excessive as 80 h—that these speeds were much higher than expected [31]. He also used spectroscopy

Fig. 3.12 Attendees at the 17th meeting of the American Astronomical Society, held August 25–28, 1914 at Northwestern University, Evanston, IL, USA. Reproduced with permission. Credit: The Hanna Holborn Gray Special Collections Research Center, University of Chicago Library

Fig. 3.13 Attendees at the 17th meeting of the American Astronomical Society, held August 25–28, 1914 at Northwestern University, Evanston, IL, USA. Reproduced with permission. Credit: The Hanna Holborn Gray Special Collections Research Center, University of Chicago Library

to investigate the rotation periods of the planets and the composition of the planetary atmospheres. As I said, in 1912, he was the first to observe the displacement of the spectral lines of the galaxies, thus becoming the discoverer of galactic redshifts (research I will elaborate on later). In 1914, he made the first discovery of the rotation of spiral galaxies [32]. In addition, he correctly measured the rotation periods of various planets in the Solar System. From his study of the spectra of the four giant planets, Jupiter, Saturn, Uranus, and Neptune, he managed to deduce the chemical compositions of their atmospheres. And, analyzing the interstellar vacuum, he found traces of sodium and calcium, which allowed him to determine the existence of large clouds of gas and cosmic dust (Fig. 3.14).

In his capacity as guardian of Percival Lowell's legacy, Slipher worked diligently in search of new indications of the existence and position of a ninth planet, beyond Neptune, something Lowell had predicted. Finally, in 1930, his efforts allowed another important observatory scientist, Clyde Tombaugh, hired by Slipher himself, to announce the discovery of Pluto. To complete this brief summary of his activities, it should be noted that, together with James Keeler (at Lick and Allegheny Observatories) and William Campbell (at Lick), they were the first to understand the importance of redshift measures, long before Edwin Hubble and Milton Humason. His cosmic observations provided the first empirical data on which the theory of the expanding

Fig. 3.14 Hot stars burn brightly in this image from NASA's Galaxy Evolution Explorer, showing the ultraviolet side of this familiar object. At approximately 2.5 million light-years away, the Andromeda galaxy, or M31, is the Milky Way's largest galactic neighbor. The entire galaxy spans 260,000 light-years—a distance so great that 11 different image segments had to be stitched together to produce this view of it. The bands of blue-white making up the galaxy's striking rings are neighborhoods that harbor hot, young, massive stars. Dark blue-grey lanes of cooler dust show up starkly against these bright rings, tracing the regions where star formation is currently taking place in dense cloudy cocoons. Eventually, these dusty lanes will be blown away by strong stellar winds, when the forming stars ignite nuclear fusion in their cores. Meanwhile, the central orange-white ball reveals a congregation of cooler, old stars that formed long ago. NASA/JPL-Caltech—NASA. Created: 15 May 2012

universe was going to be based (I will elaborate on this point below). In 1935, he was awarded the prestigious Bruce Medal of the Astronomical Society of the Pacific (ASP). Following his retirement, Slipher continued to live peacefully in Flagstaff until his death in 1969, three days before his 94th birthday. He is buried in the Citizens' Cemetery (Fig. 3.15).

Since his arrival at the Lowell Observatory, Slipher had specialized in astronomical spectrography (the branch of astrophysics that deals with the recording of the spectrum of light radiated or reflected by a celestial body). He was rather meticulous and his measurements were very accurate. He would not disclose a result before he was completely sure of it. One of his contributions in this regard was to find a way to measure the rotation speed of a planet or natural satellite around its axis. As Slipher demonstrated, this speed

Fig. 3.15 Slipher House. Author: Chris Light. Date: 7 March 2020. CC-BY-SA-4.0

can be determined by measuring the shift in the spectrum of light from the two diametrically opposite ends of the observed object (since one of these ends moves away and the other approaches us). Using this spectrographic procedure, Slipher calculated, as we have said, the rotational speeds of many objects in the Solar System.

And encouraged precisely by the positive reaction to his first studies, Slipher began, from 1912, to apply the same method to the measurement of the radial velocities of nebulae. One of the most widely discussed topics at that time was the nature of the spiral nebulae, as we have already mentioned. Until then, telescopes had not been able to reveal many details about their internal structure. At first glance, they seemed to be clouds of dust and gas, but their light had characteristics similar to the light from stars, although it was not possible to actually make out any stars. Astronomers were quite puzzled. Many believed that these were planetary systems in formation, while some believed that they could in fact be star clusters located far away, beyond the Milky Way; and that they perhaps constituted other worlds, as imagined by Kant and other thinkers.

It was in precisely the same year of 1912 in which Leavitt had published her crucial results that Slipher began his project to obtain the radial speeds of spiral nebulae from their spectra, displaced towards blue or red (because of the optical Doppler effect, already mentioned), using the 24-inch telescope at the Lowell Observatory in Flagstaff, to which Peebles refers. And it was precisely at this point that, without possible discussion, the first cosmological revolution of the last century had its origin. Slipher's first estimate, dated

September 17, 1912, was for the Andromeda Nebula, which is the closest to Earth, among the large nebulae.

Although Slipher had been an expert in handling the spectrograph since 1909, when he now turned his attention to the spiral nebulae, he found that their light was so dim that he had to increase its sensitivity, at the cost of reducing the size of the photographic plate to that of a thumbnail. This forced him to use a microscope to analyze it. Between November and December 1912, Slipher made several more observations, after which he reached a really extraordinary conclusion: Andromeda's light was shifted towards the blue part of the spectrum and, according to his calculations, this nebula was approaching us at the incredible speed of 300 km/s! This was an enormous velocity, ten times the calculated average velocity for the stars in our galaxy. So much so, that Slipher himself doubted the result could be right. But Lowell quickly encouraged him to look at more of the spiral nebulae. Next up was the Hat Nebula (NGC 4594), in the constellation of Virgo. In this case, Slipher got an even higher speed of 1,000 km per second, and this time it was moving away from us! Over the months, he began to accumulate data from several other spiral nebulae. By mid-1914, he saw that a trend was beginning to show up: most of the nebulae were moving away from us, very few were getting closer. Andromeda and other nearby nebulae were exceptions.

At the 17th meeting of the American Astronomical Society held in August 1914 at Northwestern University in Evanston, Illinois, Slipher presented the results obtained from two years of hard work, in which he had managed to measure the speeds of up to 15 spiral nebulae. The average speed he found was 400 km per second; and only three of the nebulae were approaching, the rest were moving away. His presentation was very clear and convincing and, according to the chronicles, his results were received by the audience with a standing ovation [33]. This is very unusual at a scientific conference, both then and now, and that date has rightly been gone down as an important one in the history of astronomy. Among the public, a young astronomer, Edwin Hubble, was impressed by those results, as he would much later confess, towards the end of his life. Slipher was the first astronomer to obtain spectra of galaxies with a sufficient signal-to-noise ratio to reliably measure these Doppler shifts. And his results shook the very foundations of the hitherto accepted model of a static universe.

Slipher continued to accumulate data and, by 1917, he had the spectra for 25 spiral nebulae. The data revealed that three small systems and Andromeda (all of them relatively close objects) were approaching us, while 21 more distant objects were clearly moving away. In view of his results, Slipher observed that:

This might suggest that the spiral nebulae are dispersing, but their distribution in the sky is not in accordance with this as they show a tendency to cluster.

The term "dispersion" already denotes an inclination to retreat in all directions, which could have led to the idea of an expanding universe. And added that:

> ... our entire star system moves and takes us with it. It has long been suggested that spiral nebulae are star systems seen at great distances ... This theory seems to me to be gaining importance in the light of the present observations.

In other words, Slipher correctly deduced that our galaxy was moving at a very high speed and that, most likely, the receding nebulae could be analogous to the Milky Way. This was written eight years before Hubble's detection of the famous Cepheid in Andromeda (Fig. 3.30), which finally confirmed the Kantian "island universe" hypothesis (Fig. 3.16).

In 1921 he added 13 more spiral nebulae to his speed list. Among them was NGC 584, in the constellation Cetus, which was moving away at the incredible speed of 1800 km per second, making it the fastest moving celestial object discovered up to that moment. The enormous speeds of the nebulae

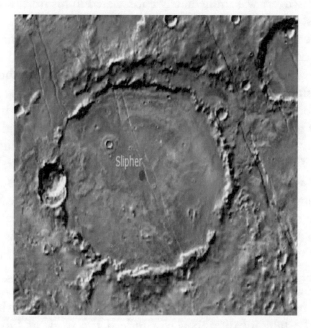

Fig. 3.16 Slipher crater on Mars (Image IAU/USGS/Goddard/A SU/USGS). Created: 17 November 2010

made it very difficult for them to belong to the Milky Way, since the gravitational field of our galaxy would hardly be able to retain them. By then, Slipher was already beginning to think that the spiral nebulae could be other universes, located beyond the Milky Way. Arthur Eddington included Slipher's velocity table in his famous book *The Mathematical Theory of Relativity* (1923), in Chapter V, p. 162. In fact, Eddington had made a special effort to include a complete list of Slipher's redshifts in his book. He obtained them from Slipher himself through direct correspondence, and they consisted of 41 speeds, with only 5 blueshifts. By then, other astronomers, such as Carl Wirtz, had already confirmed various Slipher measurements. Eddington went on to explicitly state in his book that: "*one of the most surprising problems in cosmogony is the high speed of spiral nebulae.*"

As Hubble himself would later admit, Slipher was the first astronomer to point out that something very remarkable and very strange was happening in the cosmos: how could it be static and so limited in extent with those distant nebulae escaping at such colossal speeds? Of course, peculiar speeds also came into play and one had to take into account the dipole effect that later led to the possibility of measuring the translation speed of our own galaxy. However, even in those first results, the general tendency for dispersion of distant objects was evident. It is thus clear why the great importance of Slipher's discovery was immediately appreciated in his time, even if only by a few experts, who still had (understandably) many difficulties in grasping its profound meaning and consequences for any model of the Universe, as will be further elucidated in what follows. It should be noted that there was great interest in redshift measurements generally, and their importance in contradicting the static universe model was also understood, as early as 1917, by other astronomers working at different observatories, not only by Slipher (at Lowell), but also by James Keeler (in Lick and Allegheny) and by William Campbell (in Lick), as noted above. In no sense was it an exclusive result of the Mount Wilson astronomers who incorporated them several years later. But Hubble was always very selective in his article references and did not often mention his colleagues in his publications; even Harlow Shapley, one of the contestants of the 'Great Debate' (as we will soon see) had a long dispute with Hubble on these matters. And it is precisely this that confers even greater value to the few, albeit very positive comments Hubble made about the importance of Slipher's work (Fig. 3.17).

Slipher's table of redshifts was one of the two ingredients required for Hubble's formulation of the speed vs radial distance relationship in 1929. The other, the table of distances, was indeed the work of Edwin Hubble, with a subsequent contribution from Milton Humason, who obtained some

Fig. 3.17 Vesto Slipher's spectrograph, which he used to measure the expanding universe. Slipher also used spectroscopy to investigate the rotation periods of planets and the composition of planetary atmospheres. In 1912, he was the first to observe the shifts in the spectral lines of galaxies, making him the discoverer of galactic redshifts. Author: brewbooks from near Seattle, USA. Created: 19 May 2011. CC BY-SA 2.0

additional redshifts. They were used by Hubble in 1931 to improve their earlier values. Although Milton Humason extended the spectral calculation to weaker galaxies, commissioned by Edwin Hubble, the astronomers at Mount Wilson would not have advanced as quickly without Slipher's pioneering results. Hubble was fully aware of the importance and priority of Slipher's early spectroscopy, but consistent with his style of claiming exclusive credit for most of the subjects he worked on, he never wanted to emphasize this point, which he would only recognize at the end of his life, referring to "your speeds and my distances," in a letter to Slipher of March 6, 1953. Hubble died in September of that year. In his famous book *The Realm of the Nebulae*, he clearly states, referring to Slipher:

> … the first steps in a new field are the most difficult and important. Once the barrier is overcome, further progress is relatively easy.

Now, with distances and speeds in hand, the two necessary tools were ready for the greatest revolution in the history of the study of the cosmos, which was about to take place: a conceptual change so radical that it marked the beginning of modern cosmology. But, shortly before that, there was a great debate (Fig. 3.18).

Fig. 3.18 Lick Observatory, in San Jose, California. Founded in 1888. A photochrome postcard published by the Detroit Photographic Company. Unknown author—Beinecke Rare Book & Manuscript Library, Yale University

3.4 The Great Debate of 1920: "The Scale of the Universe"

An extremely precise description of the state of knowledge in astronomy in the year 1920 can be found in these references [34], especially the last one, by the astronomer Virginia Trimble, from which I have extracted a large part of what I am going to say here, and which also contains a very detailed explanation of all the political, social, and cultural circumstances of the time in which this famous debate took place. The perspective given to us today by the century that has just passed allows us to assess the importance and significance of this anniversary. It was impossible then, and is still so today (after a hundred long years), to find a topic of greater relevance in cosmology than the one that gave the debate its title: "*On the scale of the Universe*". This fact alone, to have identified the most important possible topic for discussion, made it a great success from the outset (Fig. 3.19).

The promoter of the debate was the astronomer George Hale, son of the millionaire William Hale, who had made his fortune building elevators for the Eiffel Tower and other tall buildings. It was he who paid for his son's first telescopes, as well as for much of the very large 60-inch Mt. Wilson

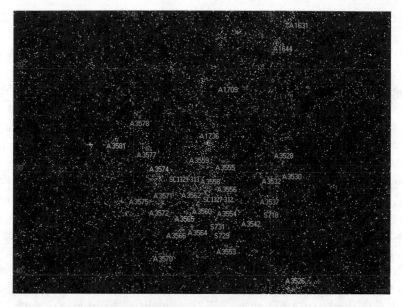

Fig. 3.19 Map of the Shapley Supercluster. Author: Richard Powell. Created: 14 March 2009. CC BY 2.5

telescope. George Hale continued this endeavour, by contributing significantly to the founding of the Mt. Palomar Observatory (in San Diego, California) and the Yerkes Observatory of the University of Chicago, and years later, he also had a lot to do with the creation of the National Research Council. It turns out that, at that time, his father had donated a fund to the US National Academy of Sciences (NAS). The money was used, among other things, to organize conferences once a year. In late 1919, George Hale convinced NAS Secretary Charles Abbot to organize a one-day conference in April of the following year in honor of his father, the benefactor William Hale. He proposed that it could be on a topic of general relativity or on the distance scale of the universe. It would be in the form of a debate, with two opposing positions. There were some discussions, not always smooth, since Abbot considered these topics to be rather specialized. In fact, Abbot would have preferred a debate on the great ice ages of the Earth or on some other popular topic of zoology or geology. Finally, they reached an agreement, both on the subject and on the opponents. Hale first proposed William Campbell, then director of the famous Lick Observatory (near San Jose, California), and Harlow Shapley, associate astronomer of Hale himself, at Mt. Wilson's. The subject of debate would be the calibration of the scale at very great distances, based on the variable stars detected in the globular clusters, along with other related considerations. It was later agreed that Heber Curtis, also from the

Lick Observatory, would replace his director; and the subject was defined as "On the distance scale of the Universe".

Hale sent the invitation telegrams on February 18, 1920 (Shapley's invitation is still preserved, Fig. 3.6). Each lecturer would receive $ 150 as fees, though the not insignificant expenses of the travel from California to Washington, which is where the debate was to take place, were to be from their own pocket. The date would be April 26, 1920, in the incomparable setting of the Baird Auditorium (Fig. 3.11) of the impressive Museum of Natural History at the Smithsonian Institution (Fig. 3.12).

The speakers agreed to discuss first among themselves the points they were going to talk about, as well as the procedure: in the morning they would separately present, one after the other, their points of view on the topic of the debate and the afternoon would be devoted to a general open discussion. Shapley would be the first to expose (Figs. 3.20 and 3.21).

Let me now make a brief presentation of the contenders. During his career as an astronomer, Curtis had become very interested in the study of the Sun, and in fact, between 1900 and 1932, he participated in eleven expeditions to observe eclipses. However, his years at Lick had been primarily devoted to photographing spiral nebulae, and it was because of this work that he was asked to confront Shapley in the debate. Shapley was the director of some observatory for much of his life, rather than a telescope astronomer. He was the one who took the reins of the Harvard College Observatory as Pickering's successor, shortly after the 1920 debate. It is said that he brought Harvard, with a steady hand, into the twentieth century; and that his work was the

Fig. 3.20 Portrait of Heber Curtis (1872–1942) cropped from a picture where he poses before the Crossley telescope. *Source* WP:NFCC#4. Uploaded: 6 November 2013

Fig. 3.21 Portrait of Harlow Shapley (1885–1972) cropped from SIA Acc. 90–105. Created: 21 April 2012

beginning of what this institution now represents worldwide. In the postwar years, he was president of the American Astronomical Society and also the American Association for the Advancement of Science (Fig. 3.22).

We should not ignore the fact that several other astronomers also contributed very relevant data and ideas to the "great debate" and, more generally, to astronomy at that time. I have already mentioned, in sufficient detail, the great contributions of Henrietta Leavitt and Vesto Slipher, but we must speak also of Johannes Kapteyn (1851–1922), who was a lifelong advocate of a small Milky Way, with the Sun at its center. The then enormously famous "Kapteyn universe," which for years was the quintessential model (what we would now call the "standard model"), long prevented attempts to even imagine the correct large-scale distribution of stars and nebulae. His tremendous influence remained for decades after his death. Adriaan van Maanen (1884–1946), for his part, was responsible for most of the measurements of the rotation of the spiral nebulae. He had first obtained the rotation of the M101 nebula or pinwheel galaxy, in 1916, and half a dozen others in the early 1920s. He got very large values (which were later shown to be completely wrong!), and this long prevented Shapley (and other astronomers) from even considering the possibility that these objects were at great distances. For, in order to have such high rotation speeds, the nebulae had to be relatively small objects and not too far away, since otherwise their

Fig. 3.22 The galaxy Messier 101 (M101, also known as NGC 5457 and nicknamed the Pinwheel Galaxy) lies in the northern circumpolar constellation, Ursa Major (the Great Bear), at a distance of about 21 million light-years from Earth. This is one of the largest and most detailed photos of a spiral galaxy released by Hubble. The galaxy's portrait is actually composed of 51 individual Hubble exposures, in addition to elements from ground-based photos. ESA/Hubble. Created: 28 February 2006. CC BY 4.0

peripheral regions would be moving at speeds greater than the speed of light (that was the argument). His device on Mt. Wilson carried the notice: "Do not use it without consulting A. van Maanen" (Figs. 3.23 and 3.24).

But Knut Lundmark (1889–1958), a visiting astronomer from Sweden, did use it without permission, a few years after the debate, to re-measure van Maanen's plates. He found no rotation, and although this point is now completely clear to all astronomers, no one has yet been able to explain precisely what van Maanen had done wrong. Moreover, there is something else here: some astronomers realized around 1924 that van Maanen's rotations were going in the opposite direction to the way they were coiled, that is, the spirals were uncoiling! Contrary to what had been correctly calculated many years before by Slipher using spectrography and the Doppler effect. It was all complete nonsense. Lundmark was, by the way, the first astronomer to note, in 1920, that some novae might be so bright as to be detectable even if they were millions of light-years from us. That debunked one of Shapley's arguments during the debate. However, he wrongly advocated a quadratic

Fig. 3.23 Knut Lundmark (1889–1958), Professor of Astronomy at Lund's University. Author: Per Bagge, 1929

Fig. 3.24 View of the newly completed Baird Auditorium, looking towards the stage, in the new National Museum Building, now known as the National Museum of Natural History. The auditorium is located under the Rotunda. The elegant classically inspired room features a domed ceiling of Guastavino tiles. National Museum of Natural History (U.S.), Natural History Building, United States National Museum. Author unknown. Historic Images of the Smithsonian. Original negative number is 97-3050. Reprinted with permission from Smithsonian Institution Archives

relationship between redshift and distance, as expected in a de Sitter universe (and as almost all astronomers of the time also expected), before Hubble and Lemaître promulgated the linear law that now bears their names. But I will discuss this later.

Returning to the debate, Shapley and Curtis disagreed on at least fourteen points, which Trimble details in Ref. [34]. They both tried, with vehemence, to impose their points of view during their morning presentations, although with very different, well-defined styles. Direct witnesses commented that Curtis was a much more experienced public speaker. He had extensive teaching experience and had prepared himself thoroughly, while Shapley had always been quite reticent when it came to lecturing. Curtis was much better organized and used slides to display graphs and other images, while Shapley limited himself to reciting from his notes. Shapley defended the orthodox view that the Milky Way was the entire Universe, while Curtis raised serious questions about this. One of Curtis' arguments was that an unusual concentration of nova stars had been found in Andromeda, pointing to the possibility that this spiral nebula was itself another world, another universe disconnected from us and similar to the Milky Way. In any case, the chronicles agree that nobody won the debate. Neither astronomer was able to convince his opponent or the audience with his arguments. And, at the end of the day, the result of the scientific discussion was a draw (Fig. 3.25).

The actors and attendees at the debate continued to speak and write about it using inflated adjectives, such as the "famous debate," "memorable session," or "anthological discussion," for several years after it took place. However, the fact is that the event seems to have attracted very little attention from the press, not even from the scientific press of the time. After a careful search, historians have been unable to locate more than a single contemporary report, dated December 1921. Moreover, in the obituary after Curtis's death, the event is not even mentioned. Shapley, in his 1969 autobiography, stated that he had long since lost all but a bare recollection of the facts. The first commemoration of the event dates from 1951, and it was not until 1960 that Otto Struve, writing on the 40th anniversary of the session, referred to it as "a historical debate," although it appears that Struve could not have witnessed the original discussions. It was in 1995, on the 75th anniversary, that a solemn diamond commemoration was organized, with all due honors, in the same beautiful and cozy place: the Baird Auditorium. In addition, the euphoria of the moment led, almost immediately, to two more debates, in 1996 and 1998[28]. Today, however, that wave has already passed, occupied as astronomers and cosmologists are right now in identifying what is finally the true value of Hubble's 'constant', the rate of expansion of the Universe (more

Fig. 3.25   National Museum of Natural History, Smithsonian Institution, Washington D.C., USA. Author: Melizabethi123. Created: 2 April 2009. CC BY-SA 3.0

on that later). In fact, an international debate on this last, crucial issue, would probably be premature at the moment (although different special sessions on it have been actually arranged in several important conferences, as the Marcel Grossmann meeting MG16 and others, during the last couple of years), and anyway, the pandemic that the whole world is still suffering from right now prevented us from celebrating this important anniversary last year in the way it deserved (the corresponding commemoration should have taken place as I was originally writing this very paragraph).

Regarding the final result of the great debate, each opponent expressed at some point the opinion that he believed he had won it. Which should not be surprising, since neither of them was clearly defeated. It was exactly like in a boxing match decided on points. The question of what is the correct scale of distances inside and outside of the Milky Way has been a subject of permanent debate since 1920, and continues to be so today, as I am writing these lines. Repeating what I have said several times, it is maybe the most difficult problem in astronomy. And the Cepheids, the standard candles *par excellence* in that debate, only constituted part of the solution, since the measurements obtained with them were far from being as precise as it was believed at the time. Suffice it to say that, using these values, Hubble and Lemaître independently made errors of between 800 and 1000%, although no one was aware of this at the time. According to the most recent results, the Shapley galaxy was too large and the Curtis galaxy too small (which is easily understood, since they were both pushing their conclusions towards opposite extremes). Shapley positioned the Sun more correctly, outside the center of the Milky Way, while Curtis still placed it in a central position. On the question of the existence of separate external galaxies, or island universes (from Kant and other writers), Curtis did make a truly important point in defending such a hypothesis. However, he himself admitted that this was only a possibility because, in his own words, "data were missing" and he could not convince his opponent or the audience on this issue. Shapley, by contrast, supported his claims that the Milky Way was the entire universe by reference to the astronomical observation of a nova in the Andromeda nebula, which had been seen to be able to eclipse, for an instant, the entire nebula. This seemed to be a completely impossible emission of energy for a single star, if Andromeda were so far away, outside the galaxy.

We have seen before that Lundmark was the first to clarify this last point, slightly later. But the full issue was definitively settled by Öpik and Hubble just a couple of years afterwards, by clearly placing Andromeda outside of the Milky Way, even outside the Shapley-inflated galaxy! His reaction to a letter that Hubble wrote to him, where he communicated his results, was recounted in the first person by a Harvard doctoral student, Cecilia Payne-Gaposchkin, who was in his office when Shapley received the letter. Addressing her (as she later testified) and while brandishing the letter in his hands, he said: *"This is the letter that has destroyed my universe."* And then, he added: *"I believed too much in Maanen's results … after all, he was my friend."* The student learned a lesson that she would never forget.

I am writing these paragraphs on April 26, 2020, precisely when the 100th anniversary of the "Great Debate" is over; and when I go back to that day, I cannot help myself getting goosebumps at times. This leads me to dedicate the following reflection to that celebration, as a small personal tribute. The perspective that the last century has since given us allows us to assess the importance and significance of that debate. It was then, and it is still today (as we will see in future chapters), impossible to find a topic of greater relevance in cosmology than the one that gave the title to that debate. This point was already an undoubted first success. In fact, the calculation of distances by Cepheids was to be the origin (see later) of the first great cosmological revolution of the twentieth century. Not just the Earth, not even the entire Solar System, but our whole Galaxy —which was the entire world, "the Universe" in full, everything known a hundred years ago—was going to become, overnight, a tiny, insignificant dot within a new Universe with no center and millions of times larger.

In the same way that, a century ago, with the knowledge that they already had of the Milky Way, astronomers could have laughed out loud at the universe of Anaximander—which placed the stars closer than the Sun—now we could also ridicule Kapteyn's model and van Maanen's, then adopted by all astronomers with very few exceptions, as the official models of their time. But we should never do this; we should never deride them: each model belongs to the age in which it was conceived, to the data and theories that were available at the time. The real lesson to learn is that our current, magnificent models—adjusted by the brightest minds to the marvelous, apparently insurmountable theories we now have—will not be the definitive ones either. And we may be sure that our knowledge of the Universe will have changed radically once again in a hundred years, when our descendants celebrate the 200th anniversary of the "Great Debate". A second lesson, no less important, is that absolute honesty and rigor should dominate data analysis. One should never 'stretch' data, as if they were made of rubber, so that they agree at all costs with preconceived estimates or ideas (for whatever reasons, even if they are very good and well intentioned), since this behavior will in the end serve only to feed a vicious circle, closing in on itself and blocking any progress. This was precisely what happened at that time, and to some extent, it continues to be repeated today, as I can testify.

A century ago, it was Leavitt's Cepheids and Slipher's redshifts which managed to break that vicious circle—a *Teufelskreis* full of erroneous results but which 'fit perfectly' with each other and which no one could even dare to question—and widen the Universe to gigantic proportions, unimaginable to the human mind. And I say the latter literally, since the cosmological revolution went (as we shall now learn) far beyond what any great thinker, outlandish visionary, or science fiction writer had ever been able to imagine. Here is a magnificent example of the theory that one particular great thinker, E. O. Wilson, defends in his excellent book "The origins of creativity" [35]. He affirms, contrary to what is generally believed, that the discoveries of science sometimes go far beyond anything that anyone could ever predict or even imagine.

3.5   An Island Universe

November 23, 1924 is the next important date in this story. That day, on the sixth page (top right-hand side) of the famous newspaper *The New York Times*, the following news item appeared:

> FINDS SPIRAL NEBULAE ARE STELLAR SYSTEMS; Dr. Hubbell Confirms View That They are 'Island Universes' Similar To Our Own.
>
> WASHINGTON, Nov. 22. Confirmation of the view that the spiral nebulae, which appear in the heavens as whirling clouds, are in reality distant stellar systems, or "island universes," has been obtained by Dr. Edwin Hubbell of the Carnegie Institution's Mount Wilson observatory, through investigations carried out with the observatory's powerful telescopes.

The first thing that jumps out here is that Hubble's name has been systematically misspelled. As much as I have tried, I have not been able to find out the true reason for this well known fact. If Hubble had sent a telegram to communicate the news to the newspaper, as would seem highly plausible at that time, it is difficult to understand the error in the transcription of his name. If he used the phone instead, as some colleagues venture, then the error is perfectly understandable, since we all know how convoluted the phonetic transcription is in English (even for the natives speakers of the language). But this is only a hypothesis. In fact, the spelling that appeared in the NYT exists and is widespread. Without going any further, the magic of the web (created by CERN and MIT scientists and engineers) has allowed me, in a matter of seconds, to find out that the owners of the Hubbell mansion in Mantorville, MN, announced on *Tripadvisor*, are fed up with the many customers who mistakenly spell it "Hubble." (Figs. 3.26, 3.27 and 3.28).

Another surprising thing, and one that I will elaborate on later, is that Hubble sent his results to the press before commenting on them to anyone; that is, before discussing them with fellow astronomers or sending them to a specialized journal to be published. Apparently, he was absolutely convinced that they would be dismissed by his colleagues and that they would be rejected if sent for publication. Personally, I find this quite logical, in view of the panorama of astronomy at that time, which I have described in sufficient detail in the preceding paragraphs. Hubble simply did not feel in any way capable of breaking the *Teufelskreis* of which I have spoken before. Despite having been hired in 1919 by Hale, an important astronomer, with all pronouncements in favor, to work at the all-important Mt. Wilson Observatory in Pasadena, California, which housed the largest telescopes in the world—as is clearly stated in the newspaper, the 60-inch, which had been

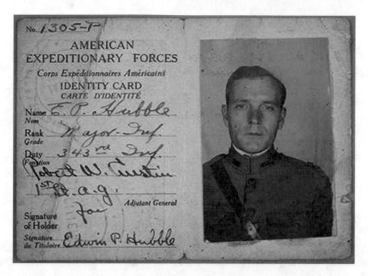

Fig. 3.26 Edwin Hubble's identity card in the American Expeditionary Forces of World War I. American Expeditionary Forces. Created: 1 January 1918

Fig. 3.27 Studio Portrait of Edwin Hubble. Photographer: Johan Hagemeyer, Camera Portraits Carmel. Photograph signed by photographer, dated 1931

there for some time, and a newly acquired 100-inch—Hubble was still a relatively young astronomer (Fig. 3.29).

A year before the day of the news in the NYT, towards the end of 1923, Edwin Hubble had focused the 100-inch telescope on Andromeda, looking for nova stars. In October, he took several photographic plates of what appeared to be three novae, which he marked with an N. To see if they

Fig. 3.28 Screenshot of the movie "Hubble—15 years of Discovery". ESA—http://www.spacetelescope.org/videos/?search=%2215+years+of+discovery%22. Created: 9 July 2013

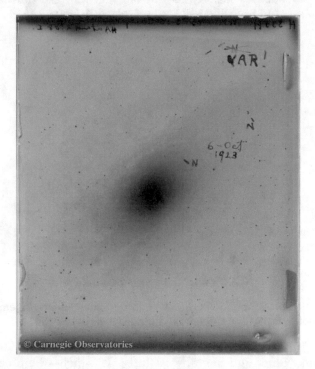

Fig. 3.29 The H335H plate that Hubble obtained on the night of 5–6 October 1923, from M31 Cepheid (Andromeda). Courtesy Carnegie Institution for Science, permission granted

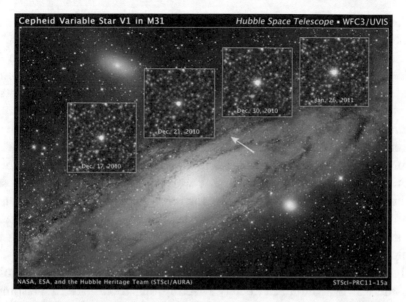

Cepheid Variable Star V1 in M31   *Hubble Space Telescope* • WFC3/UVIS

Dec. 17, 2010

Dec. 21, 2010

Dec. 30, 2010

Jan. 26, 2011

NASA, ESA, and the Hubble Heritage Team (STScI/AURA)   STScI-PRC11-15a

Fig. 3.30 NASA's Hubble Space Telescope shows the changes in luminosity of the Cepheid variable star V1 which in 1923 altered the course of modern astronomy. Credit: NASA, ESA, and the Hubble Heritage team (STScI/AURA)

actually were such, he examined older plates in great detail, with the idea of finding out when each star had first appeared. Two of them did indeed disappear when he examined old plates, but the third unfailingly remained in all plates, although its brightness varied between the different plates: it was sometimes brighter, sometimes fainter, then growing brighter again. Hubble realized then that this star was not a nova but something much more precious still, in view of a distance calculation: it was a Cepheid variable star! Immediately, Hubble crossed out, with an X, the N that he had marked on the corresponding plate, and wrote down "Var!" next to it [36]. He had just discovered the variable star which is now called V1 and which later became, quite rightly, extraordinarily famous (Fig. 3.30).

However, calculating its variability period was no easy task. Hubble had to examine a pile of old plates, dating back to 1909. On October 23, 1923, he was finally able to determine the period, 31.4 days; a very large value that had the fortunate consequence that the star had to be very distant. Using Leavitt's law he was able to deduce that the Cepheid was at a distance of 285 kpc. Therefore, that had to be the distance to the nebula that housed it, namely Andromeda. To his amazement, this distance was about ten times further than any previously calculated for celestial objects in the universe, the Milky Way, as it was understood back then. This very clearly supported the hypothesis that Andromeda was well beyond the Galaxy. It took him a further thirteen months to send the finding to the NYT.

It is interesting to note the fact that, according to Ken Croswell in *"The Universe at Midnight,"* [37] the discovery of the Cepheid in Andromeda could have been made more than ten years earlier, since, as we have seen, the photographic plates from the Mt. Wilson Observatory which Hubble examined to confirm his great find dated back to 1909. Therefore, they are almost contemporary with Leavitt's discovery, and this gave them enormous importance for calculating distances using the law that she had obtained at that time. It turns out that, by 1920, Humason had already marked possible Cepheids on plates taken by Shapley for further analysis. But Humason was then still a mere assistant, and Shapley simply erased the marks, reasoning that those stars could not possibly be Cepheids. Any further comments would be superfluous, but we would do well to take note of this new lesson.

Hubble finally agreed, although rather reluctantly, to present his results at the American Astronomical Society Meeting that took place in Washington one month after the news in the NYT (it began on December 30, 1924). His contribution was presented on January 1, 1925. In fact, it was read by Henry Russell, director of the Observatory of Princeton University. He was the one who had almost forced Hubble to report his results. He had also urged him to write them down in detail, in article form, but Hubble had not yet finished this job by then. The title of the communication was *"Cepheid Variables in Spiral Nebulae"* and was read at the joint morning session of astronomers, physicists, and mathematicians. For this, Hubble received the $ 1,000 annual award from the American Association for the Advancement of Science. They say that it was Russell himself who had also urged Hubble to apply for the award, already giving him much hope in advance that he would eventually win it.

This was a very surprising find for the time and crucial for the history of cosmology (remember Shapley's angry reaction; if he had been Japanese, it is more than likely that he would have committed *seppuku*). This was the discovery that provided a brilliant culmination to the Great Debate. Some think that, thanks to this discovery, Curtis eventually turned out to be the virtual winner of that debate. It also provided a foundation for previous ideas expressed by philosophers, like Immanuel Kant (in 1755) [38], and by some writers, including Edgar Allan Poe (see his "Mesmeric Revelation," from 1844), who had already advocated in their works that the universe was probably much larger than the Milky Way, and that it was a universe made of islands ("island universe"), each similar to our own galaxy. In fact, the conjecture of the possible existence of other universes already had a long history, dating back to at least the eighteenth century (see Michael Way [39]); even the famous astronomer William Herschel, mentioned earlier, was convinced [40] for a time that the spiral nebulae were outside the Milky Way, although he then changed his mind [41]. The vision of the cosmos switched completely as a consequence of the Hubble discovery, although not as quickly as the reader may perhaps be thinking upon reading this (Figs. 3.31 and 3.32).

Everything above has been stressed, one could even say exalted, and rightly so, countless times. What worries me, however, is the existence of another fact—also of extraordinary relevance—which Hubble does not seem to have been aware of at the time. That would be understandable, but incredibly, many scientists and almost all cosmology texts, including the vast majority of references on the subject, continue to ignore this fact even today. For me this is something quite incomprehensible. It turns out that, two years before Hubble and only two after the great debate had taken place, a young, but highly competent Estonian astronomer, Ernst Öpik, had published an article without fuss or bother in the most prestigious journal, *The Astrophysical Journal* in which he obtained a distance to Andromeda of 450 kpc, much closer to the value calculated today, of 775 kpc, than the value obtained by Hubble two years later! Öpik had used a very different method for calculating the distance, which had nothing to do with the presence of a Cepheid star, but proved to be doubly accurate in that case. His procedure was based on the use of the rotation speeds observed around the galaxy (which inform us of its mass) and on the hypothesis that the luminosity per unit mass was the

Fig. 3.31 Immanuel Kant (1724–1804). Kant portrait by Johann Gottlieb Becker. Created: 1768

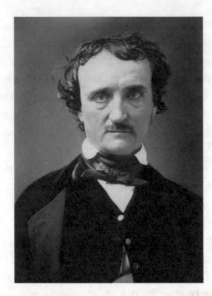

Fig. 3.32 Edgar Allan Poe (1809–1849). "Annie" daguerreotype of Poe. Unknown author. Restored by Yann Forget and Adam Cuerden. Created: 1 June 1849

same as for the Milky Way, which allowed him to infer Andromeda's true luminosity. This method was used by many other astronomers at the time and is still in use today. Öpik was already somewhat well known, having been the first to calculate, six years earlier, the density of a white dwarf. He made several other important contributions to astrophysics, too, for which he received numerous distinctions. But no one seems to have paid any attention to his very important result, in 1922, and very few, outside the community of astronomers, even know his name today (Fig. 3.33).

It is sad and disappointing to realize how much it costs to change the description of past events. And this in spite of the effort of so many historians—and lately also of ordinary mortals, too, with the magnificent tools accessible on the web—spent in bringing to light, day after day, the gross mistakes that are made in many of these descriptions. In fact, there is a theory, which I will discuss later, which states that in all of them. I confess that this has been my greatest motivation in writing this book, and it is the reason for its title. Öpik, like Slipher, was a level-headed and down-to-earth worker, who did not spend time publicizing his results and did not seek to be in the foci of influence of science; and nor did he own the largest telescope in the world. Hubble may be excused for not reading The Astrophysical Journal and thus remaining completely unaware of Öpik's result, even though it was much better than the one he himself obtained. But there is no excuse for this tremendous laziness in amending the history books and changing the stories

Fig. 3.33 Ernst Öpik (1893–1985). This image is the property of Armagh Observatory and is released on their behalf by Martin Murphy

from top to bottom, in view of the new evidence, which was clearly proven. It seems that chroniclers, out of laziness, do not like to spend time getting to the bottom of the facts, and simply limit themselves to copying what others have already written before. Anyway, it is not possible to modify the conclusion of the story in this section; this one cannot be rewritten.

Let us see why. It is an incontestable historical fact that, due to his greater influence, it was Hubble who eventually changed our vision of the universe. But this did not happen in 1924. Even being who he was, making this change cost him a lot of time and effort. The paradigm shift happened much more slowly than we might now think. Even after the publication of his results on Andromeda, which did not take place until 1926, many of his colleagues received them with enormous skepticism, or worse still, with total indifference. It is only in our current vision, with the perspective we now have of the past century, that we are able to recognize the importance of the historical discovery made by Öpik and Hubble. And it looks so obvious to us that we cannot put ourselves in the situation of a hundred years ago, when it took the scientific community decades to realize what these discoveries really meant. Below are brief biographies of Hubble and Öpik.

3.5.1 Edwin Hubble

Edwin Hubble was born in Marshfield, Missouri, in 1889, and when he was eleven years old his family moved to Wheaton, Illinois. As a young man, he stood out more for his numerous sporting achievements than for his intellectual ones, but he generally obtained good grades. Some biographies affirm that he intended to become a heavyweight boxer, and what is certain is that he led the team from the University of Chicago, where he studied, to its first title in 1907. He began by studying law, following the advice of his father, who worked as an executive in an insurance agency, although he also took Spanish, mathematics, and physics classes, and in fact ended up graduating in science in 1910. He then spent three years at Queen's College, Oxford, in England, with one of the first Rhodes scholarships of his university, and obtained there a master's degree in jurisprudence, thus finally fulfilling the promise he had made to his father, who died in 1913, while Hubble was still in England. In fact, since childhood he had been interested in astronomy, but he obeyed his father until the latter's death. Returning to the United States, instead of practicing law, he spent a year teaching Spanish, physics, and mathematics at New Albany High School in Indiana. Then he came back to the University of Chicago with the intention of obtaining a doctorate, and it was there that he first began to study astronomy, at the Yerkes Observatory in Williams Bay, Wisconsin, which belonged to the university. The observatory was inaugurated in May 1897, thanks to the astronomer George Hale who was its promoter and had convinced the president of the University of Chicago of the need to build an astronomical observatory. Funding for it was provided by magnate Charles Yerkes (1837–1905), whose body is buried at the base of the main telescope. Hubble soon found himself in his element at the observatory and obtained his doctor's degree in 1917, with the thesis "*Photographic Investigations of Faint Nebulae.*" The Yerkes Observatory, with its 24-inch reflector, had one of the best telescopes of the day. In fact, Hubble had to hurry to submit his thesis, in order to be able to enlist and serve in the Great War as a volunteer, although the division assigned to him never entered combat. He then spent a year at Cambridge University, before being recruited for a staff position at the Carnegie Institution's Mt. Wilson Observatory. He was taken on by George Hale, its director and founder in 1904. Hale, whom we have already met on several occasions, intended to emulate William Herschel, also encountered in the first chapter, and like him spent his life trying to build ever-larger telescopes. Although he had received the offer months before, Hubble did not join Mt. Wilson until 1919, upon his return from England (Figs. 3.34 and 3.35).

Fig. 3.34 The astronomer Edwin Hubble. http://recherche-technologie.wallonie.be/fr/particulier/menu/revue-athena/par-numero/numeros-anterieurs/septembre-2006-a-juin-2007/n-232-juin-2007/astronomie/index.html?PROFIL=PART. Created: 1920. Public Domain

Fig. 3.35 Milton Humason (1891–1972). Courtesy Carnegie Institution for Science, permission granted

He quickly became the main user of the powerful 100-inch Hooker telescope, which had been installed in 1917. From the beginning, he focused his attention on nebulae. He discovered that many of them contained not only gas, but also a huge number of stars. He was soon impressed by the skills and absolute dedication of Milton Humason who, although he had some scientific training, did not have a university degree and had already started working at the observatory in 1909 as a mule groom during its construction. He had then become a doorman, electrician, and night assistant, until his father-in-law (he had married the daughter of an observatory engineer) invited him to work as an observation assistant, in 1917, whereupon he became a member of the observatory's staff. Legend has it that one night the telescope operator fell ill and the astronomer on duty asked Humason if he would be able to take his position for a couple of days. He carried out the task with such skill that he soon went on to occupy a permanent position as telescope operator and astronomical assistant. Hubble suggested that he take redshift measurements of some distant nebulae, and Humason proved himself extremely capable. He specialized in spectroscopy and his particular ability to get the most out of the Hooker telescope is legendary. Indeed, in the 1930s he was able to measure the extremely faint spectra of distant galaxies. Hubble and Humason formed a perfect team for years. Humason eventually published numerous research papers and discovered a comet that now bears his name.

Aside from Andromeda, Hubble was able to determine that there were other nebulae outside the Milky Way, calling them "extragalactic nebulae" (he never liked to use the increasingly common name of "other galaxies"). That there could be other galaxies beyond our own had been suspected, as I said, for a long time, although this idea was rejected by almost all astronomers at the time, Harlow Shapley being one of the most vociferous opponents. Once they had been discovered, Hubble began classifying these extragalactic nebulae. He identified spirals, ellipticals, lenticulars, and irregulars, and defined a sequence that is still known today as "the Hubble sequence." He remained at the Mount Wilson Observatory until his death in 1953. Shortly before that, Hubble was the first astronomer to use the giant Hale 200-inch reflector telescope at the Palomar Observatory near San Diego, California, which was already mentioned in the last section.

During his lifetime, Hubble received many distinctions and recognitions, which would be too long to list here. Without a doubt, the one that has brought him most clearly into the public eye was the fact that the space telescope that NASA launched in 1990 was named after him. On April 24, 2020, as I write, Hubble has just celebrated 30 years in space. The extraordinary and striking images of the Hubble Space Telescope are known in all corners of the Earth. I would even say that these pictures reach far into the depths of the human soul. Surely, they have done more for astronomy than thousands of encyclopedias, treatises, and conferences. It is my opinion that they have also greatly influenced the fact that Edwin Hubble is considered today as one of the most important astronomers of the twentieth century. Without intending to take even one inch of the enormous value from all his contributions, in this chapter and the next I will only try to put things in what I consider to be their rightful place (to Caesar what is Caesar's …).

3.5.2 Ernst Öpik

Ernst Öpik was born in 1893, in Kunda, Estonia, which was then part of the Russian empire. Although he spent the second half of his life (between 1948 and 1981) at the Armagh Observatory in Northern Ireland. Öpik studied at the University of Moscow, specializing in observing celestial objects such as asteroids, comets, and meteorites. He completed his doctorate at the University of Tartu. In 1916 he published a pioneering article in *The Astrophysical Journal*, in which he estimated the densities of some visible binary stars, in particular, the white dwarf Eridani B, for which he obtained a density 25,000 times greater than that of the Sun. He concluded from the outset that the result was simply impossible. In 1922, he published in the same magazine the crucial work [34], in which, as already mentioned, he calculated the distance to Andromeda using a novel astrophysical method based on the observed rotational speeds of stars in the nebula. These depended on the total mass around which the stars revolved, and assuming that the luminosity per unit mass of the nebula was the same as that of the Milky Way, he concluded that the distance was 450 kpc. This result was much closer to the current estimate of 778 kpc than the one obtained two years later by Hubble (275 kpc), and it already placed Andromeda very clearly outside our Galaxy. His method is still used today. In 1922 he correctly predicted the abundance of craters on Mars, long before it could be verified by space probes. In 1932, he formulated a theory about the origins of comets in our Solar System. He postulated that they orbited in a cloud that had to lie beyond the orbit of Pluto, and that would later be known as the Oort cloud, although some call it the Öpik-Oort cloud (Fig. 3.36).

In 1944 Öpik left his native country for fear of an imminent incursion of the Red Army into Estonia. He lived in Germany as a refugee and served as rector of the Baltic University for Refugees. In 1948, the Armagh Observatory offered him a job that he did not hesitate to accept. He remained there for the rest of his days, despite having received much more lucrative offers from American universities, although it is true that, from 1960 to 1970, he combined this with a position at the University of Maryland. In 1951 he published an article on the triple alpha process, which describes the combustion of helium-4 to give carbon-12 in the nuclei of red giant stars. However, this achievement is often overlooked, because Edwin Salpeter's article on the

Fig. 3.36 Color image of Phobos, imaged by the Mars Reconnaissance Orbiter on 23 March 2008. Phobos (systematic designation: Mars I) is the innermost and larger of the two natural satellites of Mars, the other being Deimos. Both moons were discovered in 1877 by American astronomer Asaph Hall. Phobos is a small, irregularly shaped object with a mean radius of 11 km. It orbits 6,000 km from the Martian surface, closer to its primary body than any other known planetary moon. Geological features on Phobos are named after astronomers who studied this moon and people and places from Jonathan Swift's Gulliver's Travels. The crater at the center of the image is named after Ernst J. Öpik. Wikipedia. NASA/JPL-Caltech/University of Arizona

same subject had already been published when Öpik's paper became known in Britain and the United States. He was the inventor of the rocking camera, which allows us to determine the speed at which meteors enter the atmosphere, and he proposed a pulsating universe theory. He died in Bangor, Northern Ireland, in 1985.

Öpik won the J. Lawrence Smith Medal of the National Academy of Sciences, in 1960; the Meteorite Medal of the Frederick C. Leonard Memorial Society, in 1968; the Kepler Gold Medal of the American Association for the Advancement of Science and the Meteorite Society, in 1972; the Royal Astronomical Society Gold Medal, in 1975; and the Bruce Medal, in 1976. Asteroid 2099 Öpik was named in his honor, as was the Öpik crater on the Martian moon Phobos.

4

The Theory of General Relativity and Its Main Solutions

The first cosmological revolution had three important and well-differentiated milestones. The first I have just explained: the finding that the Universe was millions of times larger than the Milky Way. The second will be presented in this chapter: contrary to the prevailing view in the 1920s, supported by rigorous physical considerations that everyone, including Einstein, accepted at face value, the Universe was not static, but expanding. The third, which will be discussed in the next chapter, was to realize that our Universe had not existed forever: it had an origin, some fourteen billion years ago. A further very important point was discovering how—in accordance with the fundamental laws of physics—this origin might have occurred (what is now known as inflation or the Big Bang).

It should made quite clear that this great revolution did not take place at any specific time. We may look as hard as we like for a specific event, such as the taking of the Bastille in the French Revolution, or the publication of a founding document, like Darwin's "The Origin of Species," or "*De revolutionibus*," which marked the Copernican Revolution, but we will not find any. The cosmological revolution to which I refer extends over more than half, or perhaps two thirds, of the twentieth century. It was, in this sense, a cumulative, smooth, and continuous process, in which useful knowledge accumulated little by little, while wrong proposals (some of which seemed to have been firmly established) were dismissed one after another. Looking for a specific time, I have clearly identified the year 1912 as marking the beginning of the first milestone, for the reasons previously explained. In that year, the

© The Author(s), under exclusive license to Springer Nature
Switzerland AG 2021
E. Elizalde, *The True Story of Modern Cosmology*,
https://doi.org/10.1007/978-3-030-80654-5_4

weapons necessary to "take the Bastille"—that is, the tools required to destroy the evil cosmic *Teufelskreis* of which I have already spoken—were first made available to astronomers to initiate what in the end was going to become the first cosmological revolution of the past century.

There is still one very important aspect which will not have escaped the reader's attention. I said before that the cosmological revolution exactly coincided with the time when cosmology itself became a modern science, a true science, in all senses of the word. The tools I have just mentioned correspond only to the empirical aspect of cosmology, namely, the observation of celestial objects, including the determination of their positions and speeds, and of their evolution. But the results obtained from such observations must be given coherence. They must be *explained* on the basis of some fundamental law, which, if it is as good as it is supposed to be, must be predictive and tell us how these objects and the entire universe will evolve in the future, and how it evolved in the past; even before the astronomical data to confirm both these extremes are available. (Now please read this again.)

However, I am getting a little too far ahead of events. Back to the point, the fundamental law *par excellence* in the case at hand is none other than Albert Einstein's general theory of relativity (more commonly just called general relativity, or GR). This does not mean that modern cosmology does not also rely, and very substantially, on other laws of physics. Indeed, it depends increasingly on the other great theory produced by the second twentieth century revolution in physics, viz., quantum mechanics and, in particular, quantum field theory (QFT). However, I shall say little about these, which go beyond the scope of the present book.

In short, general relativity is the theory that has given cosmology as a science its theoretical pillar or foundation, and I will dedicate the following section to it.

4.1   Einstein's General Relativity

What could I say about Einstein that no one has yet said? Very little, in my opinion. Just change the adjectives, one here, one there, and perhaps also the emphasis, in the presentation of his great discoveries. Or maybe highlight some of his weaknesses. Either way, I am going to try to write something sufficiently original and meaningful. Albert Einstein was declared by "Time" magazine as the "*Person of the 20th Century*". You need to pay close attention: the meaning 'person' includes all the rational beings that lived during the past century. No one puts in doubt that he is also one of the greatest thinkers in the whole of human history. In 1999, his biographer, Walter Isaacson, wrote: *"The twentieth century will be remembered, above all, for the advances in science and technology"* and that is why he was not surprised by the choice made by the famous magazine. In a different context, the great mathematician Godfrey Hardy, discoverer and mentor of the genius Srinivasa Ramanujan, correctly wrote in 1940 that: *"Greek mathematics is permanent, even more so than Greek literature. Languages die, mathematical ideas survive forever."* And as though to continue along the same lines, the editorial article of Time, from 1999, stated (Fig. 4.1):

In a hundred years, in a hundred times a hundred years from now, for many millennia that pass, the name that will be more enduring in our entire era –admirable, by the way, for so many, many things– it will be Albert Einstein's.

A short biography follows, which in no way intends to be the best summary of what has been written about his life. Einstein was born on March 14, 1879, in Ulm, Germany, when it was still an empire, but his family moved to Munich a year later, and his sister Maya, two years younger, was born there. They say that Einstein was very reserved and introverted as a child, and did not learn to speak until he was three years old. This fact,

Fig. 4.1 Albert Einstein (1879–1955). Author: Vlad Grigoryan—Own work. Created: 16 April 2020. CC BY 4.0

together with his character, made his parents wonder if their child suffered from some intellectual disability. However, this did not prevent him from becoming a genius. Einstein always alleged that, if he was able to develop the theory of relativity, it was due to the very slow development of his intellect. Indeed, a normal adult never wonders about apparently banal questions, like the nature of time and space. This you can only do when you are a child. His mother, who played various musical instruments, soon inspired the passion that Einstein showed for music from a young age. He was also greatly influenced by his uncle Jakob Einstein, an engineer, who encouraged him to read science books. His father and uncle founded a company in Munich specializing in the installation of water and gas systems. As the business was going well, they decided to open their own electrical equipment workshop (*Elektrotechnische Fabrik J. Einstein & Cie.*). It supplied power plants in Germany and Italy. However, their enterprise would soon fail, thus indebting the entire family. This was a traumatic experience, forcing them to sell the beloved garden of the Munich house to a property developer to pay off debts and finance their transfer to Milan (Italy).

At the age of four, in the course of an illness that kept Albert in bed, his father gave him a pocket compass. In his own words, this event would be decisive for the little Einstein, since he was at once fascinated by the fact that the needle, without being in contact with anything, always pointed in the same direction. His innate curiosity would be encouraged by his parents, who educated him in perseverance and independence. Einstein attended primary education at a Catholic school in Munich, obtaining excellent grades, especially in science. The secondary stage was harder. In 1895 he was reunited with his family in Milan. But Einstein had not yet finished high school. He tried to get into the Swiss Federal Polytechnic Institute (Eidgenössische Polytechnische Schule, later the famous ETH) through an entrance exam, but there was one subject he did not pass in the humanities. However, the following year, he did obtain a bachelor's degree and, at the age of 17, he was able to enter the Zurich Polytechnic to study Physics (Figs. 4.2 and 4.3).

During his years in Zurich, he discovered the works of various philosophers and thinkers, like Henri Poincaré, Baruch Spinoza, David Hume, Immanuel Kant, Karl Marx, and Ernst Mach. He also made contact with the socialist movement through Friedrich Adler and with certain maverick and revolutionary thinkers, along with his longtime friend Michele Besso. In October 1896, he met Mileva Marić, a Serbian classmate of radical spirit, with whom he fell in love. In 1900, Albert and Mileva graduated from the Zurich Polytechnic. With this, he obtained the title of professor of mathematics and physics. In 1901, at the age of twenty-two, he obtained Swiss citizenship,

Fig. 4.2 Henri Poincaré (1854–1912). Public Domain

which he kept throughout his life. During this period, he discussed his scientific ideas with a group of close friends, Mileva included. He secretly had a daughter with her, in January 1902, whom they called Lieserl (Figs. 4.4 and 4.5).

Einstein was unable to find work at the University for some time, so he served as tutor in Winterthur, Schaffhausen, and Bern. His classmate Marcel Grossmann, who would later help him in the mathematical formulation of general relativity, also helped him to find a job at the Swiss Federal Intellectual Property Office, in Bern. Einstein worked there from 1902 to 1909 as a patent examiner. His personality is said to have caused him some trouble with the director, who taught him to "express himself correctly." At the time, Einstein referred to his wife Mileva with love and admiration, as "*a person who is my equal and as strong and independent as I am.*" Abram Joffe, in his biography of Einstein, argues that Mileva helped him with his research during this period. However, this contradicts other biographers, such as Ronald Clark, who affirm that Einstein and Mileva had a somewhat distant relationship, which gave him the necessary solitude to be able to fully concentrate on his work [44].

The first article by Einstein was about capillary attraction [45]. He sent it for publication to *Annalen der Physik*, in 1900. It was published in 1901, with the title "*Folgerungen aus den Capillaritätserscheinungen*" ("*Conclusions that follow from the phenomenon of capillarity*"). In two articles he published on thermodynamics in 1902 and 1903, he attempted to interpret atomic phenomena from a statistical viewpoint. They formed the basis of his famous

Fig. 4.3   Hendrik Lorentz (1853–1928) in front of a blackboard with tensor formulas from Einstein's general theory of relativity in Lorentz' handwriting

1905 work on Brownian motion, in which he showed that it can be interpreted as firm proof of the existence of molecules. His research in 1903 and 1904 focused primarily on the effect of finite atom size on diffusion phenomena (Figs. 4.6 and 4.7).

In May 1904, Einstein and Mileva had a son, whom they named Hans Albert. That same year, his job at the Patent Office was made permanent. Shortly afterwards, he completed his doctorate and presented the thesis entitled "*A new determination of molecular dimensions*" (which I had the pleasure of being the first to translate directly from German to Catalan, a few years ago). It consists of just 17 pages, and emerged from a conversation he had with Michele Besso, while they were having a cup of tea. As Einstein added sugar to his own, he asked Besso: "*Do you think calculating the dimensions of the sugar molecules could be a good PhD thesis?*" And from this short conversation the whole thesis was born. Already in his early career, Einstein had deeply

Fig. 4.4 Portrait of Albert Einstein by Emil Orlík, 1923

reflected on the fact that Newtonian mechanics did not seem sufficient to reconcile the laws of classical mechanics with those of electromagnetic fields. And that led him to develop his special theory of relativity, in his spare time, during his years at the Swiss Patent Office.

While working there, in 1905—which is now known as his *annus mirabilis* (miraculous year) —he published four innovative documents which have stunned the academic world ever since. In the first paper he presented the theory of the photoelectric effect; in the second, he explained the laws of Brownian motion; in the third, he introduced the special theory of relativity; and in the fourth, regarding the equivalence of mass and energy, he made its consequences explicit, in what is today considered to be the most famous formula ever written, $E = mc^2$. That year, at the age of 26, he received his doctorate from the University of Zurich. And these articles earned him thereafter, a position at the University of Bern in 1909; another at the Prussian Academy of Sciences in Berlin in 1914; and the Nobel Prize in Physics, in 1921, awarded for his work on the photoelectric effect (Fig. 4.8).

Although initially treated with considerable skepticism by the scientific community, Einstein's work began to get recognition little by little, when his colleagues gradually realized what extraordinary and significant progress he had made. Einstein originally conceived his theory of relativity in purely

Fig. 4.5 Portrait of Michele Besso (ETH-Bibliothek Zürich, Bildarchiv). CC BY-SA 4.0

Fig. 4.6 Albert Einstein and his first wife Mileva Marić. Credit ETH Zurich Archives. CC BY-SA 4.0

Fig. 4.7 Einstein at a picnic near Oslofjorden in connection with his visit to Norway, 20 June 1920. Ole Colbjørnsen, who was teaching the theory of relativity, invited Einstein to Oslo. From left to right, we see Heinrich Goldschmidt, Albert Einstein, Ole Colbjørnsen, Jørgen Vogt, and Ilse Einstein. *Source* MUV, University of Oslo, Norway. Author: Halvor Rosendahl

kinematic terms (the study of bodies in motion), but in 1908 Hermann Minkowski reinterpreted it geometrically, as a true theory of space–time. And this was the formalism that Einstein adopted in his general theory of relativity of 1915. Before that, in 1908, at the age of twenty-nine, he was hired by the University of Bern, Switzerland, as Privatdozent and the following year he moved to the University of Zurich. Einstein and Mileva had another son, Eduard, in 1910. Shortly after, the family moved to Prague, where Einstein obtained the position of professor of theoretical physics at the German University of Prague, having to adopt Austrian nationality to gain access to the position. After Prague, Einstein went back to Zurich (August 1912), where he began to work closely with Marcel Grossmann and Otto Stern. He also began to call the mathematical time *"the fourth dimension."* In 1913, just before the First World War, when he was elected to the Prussian Academy of Sciences, he established his residence in Berlin, where he remained for seventeen years. Emperor William invited him to direct the physics section of the Kaiser Wilhelm Institute of Physics (Figs. 4.9, 4.10 and 4.11).

Fig. 4.8 Photograph of Albert Einstein in his office at the University of Berlin, published in the USA in 1920. *Source* ETH Archiv

After publishing his work on special relativity in 1905, Einstein set about extending the theory to the gravitational field. This was not easy for him, despite the fact that the final form of the equations he obtained was very compact and elegant. It cost him, from the outset, ten whole years of his life, because it was not until 1915 that he was able to complete a final version of his general theory. As he later confessed, the reason that led him to the development of general relativity was that the preference for inertial motion within special relativity was unsatisfactory. He felt that a theory which from the outset did not prefer any particular state of motion (not even an accelerated one) would be more satisfactory [46].

In accordance with this reasoning, in 1907, he published an article on acceleration within special relativity, entitled "*On the principle of relativity and the conclusions drawn from it.*" He argued there that free fall is in fact an inertial motion, and that for a freely falling observer, the rules of special relativity should apply. This argument is the famous *equivalence principle*. In the same

Fig. 4.9 Portrait of Hermann Minkowski (1864–1909), cropped from a scan taken from Raum und Zeit (Jahresberichte der Deutschen Mathematiker-Vereinigung, Leipzig, 1909.) Date: 17 August 2017

article, Einstein obtained the important phenomena of gravitational dilation of time, gravitational redshift, and deflection of light by the gravitational field [47, 48]. In 1911, he published another article *"On the influence of gravitation on the propagation of light,"* in which he expanded this last concept, quantitatively estimating the deviation of light by massive bodies. Thus, a theoretical prediction of general relativity could be tested, for the first time, experimentally [49] (Figs. 4.12, 4.13 and 4.14).

However, only after countless corrections, revisions, and restatements was he able to arrive at the version of the general theory of relativity that is currently considered correct. According to the thesis of my student Martí Berenguer, presented in his degree thesis at the Universitat Autònoma de Barcelona (UAB) [50], the process of constructing the general theory of relativity could be divided into two quite distinct stages:

(i) A first stage that extends from 1907 to 1912, during which Einstein developed a series of physical arguments and thought experiments (Gedankenexperimente) with the aim of finding a theory that generalized special relativity (published by himself a couple of years earlier) to accelerated movements. Einstein realized at that time that accelerations

Fig. 4.10 Portrait of Marcel Grossmann (1878–1936). File: ETH-BIB-Grossmann. Created: adadfadsf

of any kind and the effect of gravitational fields on masses were essentially one and the same thing. This led him to obtain, in 1912, a theory for static gravitational fields.

(ii) The second stage covers from 1912 to 1915, years in which Einstein dedicated himself to perfecting his theory to make it more general. To do this, he needed to include more sophisticated mathematical concepts, which led him to seek the help of other colleagues who possessed such knowledge, especially Marcel Grossmann and David Hilbert. In 1912, Einstein introduced the metric tensor to characterize the gravitational field. And in 1913 he published what is often called his *Entwurf* (or draft), a first version of the theory [51]. Although not yet correct, this already provided Einstein with all the formalism needed to complete, on November 25, 1915, the definitive version of his theory: the field equations corresponding to his general theory of relativity. The article was first published in December 1915, in the *Sitzungsberichte der Preußischen Akademie der Wissenschaften zu Berlin* (the first page of this article is shown on p. 65, Fig. 88). In 1916, an article appeared in the *Annalen der Physik*, which is a summary and extension of the four articles in the

Fig. 4.11 David Hilbert (1862–1943). Photograph taken in 1912 for postcards of faculty members at the University of Göttingen, which were sold to students at the time

Sitzungsberichte, all submitted in November 1915. But it was not until the following year, 1917, that he applied the general theory of relativity to find a model of the universe. All this I will detail later.

In 1916, Einstein further predicted the existence of gravitational waves [52], ripples in the curvature of space-time that propagate as waves, traveling from their source and transporting energy in the form of gravitational radiation. The existence of such waves is possible in general relativity as a consequence of Lorentz invariance, which implies a finite speed of propagation for the gravitational interaction. They cannot exist in Newtonian gravitation, where physical interactions propagate at infinite speed. The first (indirect but very accurate) detection of gravitational waves was in the 1970s with the observation of two neutron stars in close orbit, referred to as PSR B1913 + 16. This resulted from the work of Russell Hulse and Joseph Taylor in Princeton University, New Jersey, who were awarded the Nobel Prize in 1993 [53]. The only plausible explanation for the tiny variation in their orbital period was that they emitted gravitational waves. Fifty years of observing the phenomenon have provided the most extraordinary adjustment

Fig. 4.12 Albert Einstein during a lecture in Vienna in 1921. Author: Ferdinand Schmutzer

known to date due to the predictions of general relativity. Then, the existence of gravitational waves was directly confirmed in 2016, when the LIGO researchers published the first direct observation [54], detected on Earth on September 14, 2015, almost one hundred years after being predicted by Einstein [55]. This result was also recognized with the Nobel Prize, awarded to Rainer Weiss, Barry Barish, and Kip Thorne in 2017 (Figs. 4.15, 4.16 and 4.17).

But, I have got too far ahead of myself once more. Going back a century again, the first major confirmation of Einstein's theory came as early as 1919, when a solar eclipse was photographed by an English astronomical expedition. On Fig. 4.15 we see the photo taken by Arthur Eddington from the island of Príncipe, off the coast of Africa, on May 29, 1919, to test Einstein's theory. The image confirmed that the Sun had deflected the light from the stars, since it was observed that those that were viewed near the Sun during the solar eclipse were in slightly different positions from those they would have had if the Sun had not been there. So, the Sun had diverted the light that came from them. In fact, it was noticed a posteriori that the signal contained

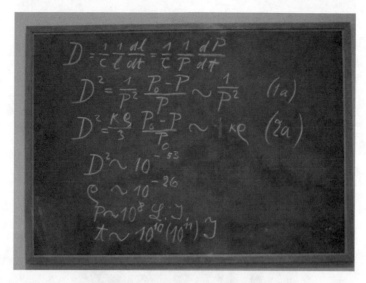

Fig. 4.13 A blackboard used by Albert Einstein in a 1931 lecture in Oxford. The last three lines give numerical values for the density, ρ, radius, P, and age of the universe. The blackboard is on permanent display in the Museum of the History of Science, Oxford. *Source* decltype—Own work. Created: 18 April 2010. CC BY-SA 3.0

Fig. 4.14 ETH Zurich University—Albert Einstein's locker (by Ank Kumar). Created: 2 September 2019. CC BY-SA 4.0

Fig. 4.15 Photograph of the 1919 eclipse taken by Eddington from the Princes' Islands. From the report of Sir Arthur Eddington on the expedition to check Albert Einstein's prediction of the bending of light around the sun. The plate shows a half-tone reproduction of one of the negatives taken with the 4-inch lens at Sobral. This shows the position of the stars, and, as far as possible in a reproduction of this kind, the character of the images, and, as there has been no retouching. Frank Watson Dyson–F. W. Dyson, A. S. Eddington, and C. Davidson (1920), "A Determination of the Deflection of Light by the Sun's Gravitational Field, from Observations Made at the Total Eclipse of May 29, 1919". Philosophical Transactions of the Royal Society A: 332. ISSN 1364-503X. Created: 29 May 1919

quite a large error bar: there were various measurements from different photographic plates with a total uncertainty of about 30%; this was confirmed later by more accurate analyses [56]. Anyway, following the compelling confirmation of his theory at that point, Time presented Einstein as the new Newton and his international recognition was immediate and enormous. On February 14 of the same year 1919, at the age of thirty-nine, he divorced Mileva, after a marriage of sixteen years, and a few months later he married his cousin, Elsa Loewenthal (Fig. 4.18).

During that time, Einstein worked on statistical mechanics and quantum physics problems. He also investigated the thermal properties of light and the quantum theory of radiation, which laid the foundations of photonics and the development of the laser decades later. In 1924, the Indian physicist Satyendra Nath Bose sent him a description of a statistical model, based on a counting method that assumed that light could be understood as a gas of indistinguishable particles. Einstein pointed out that the same Bose statistic

Fig. 4.16 Richard Tolman and Albert Einstein standing in front of the blackboard at the California Institute of Technology in 1932. Notes on negative states "Dr. Richard Tolman & Dr. Einstein". Los Angeles Times photographic archive, UCLA Library. Created: 9 January 1932

Fig. 4.17 Einstein in Espluga de Francolí (Tarragona, Spain) on February 25, 1923. Author: Casimiro Lana. Public Domain

could also be applied to some atoms—in addition to light particles, as Bose proposed—and sent his translation of Bose's article to Zeitschrift für Physik for publication. Einstein also published several articles studying the model and its implications, including the very important phenomenon known today as Bose-Einstein condensation, which certain particles should experience at

Fig. 4.18 Together with Rey Pastor, Terradas, Cabrera, Ramón y Cajal and other scientists of the time, the engineer, professor, and politician Casimiro Lana Sarrate organized the arrival of Albert Einstein in Spain, in 1923. Credit: Casimiro Lana Sarrate. Archivo General de la Administración [IDD(03)88, F/03198, S10, F16]

very low temperatures [57]. It was not until 70 years later, in 1995, that the first condensate of this type was produced experimentally by Eric Cornell and Carl Wieman using ultra-cooling equipment built in the NIST-JILA laboratory at the University of Colorado at Boulder. Cornell and Wieman received the NP in 2001 [58]. Bose–Einstein statistics is now used to describe the behavior of any set of bosons. In 1933, while Einstein was visiting the United States, Adolf Hitler came to power. Being of Jewish origin, Einstein did not return. He became a U.S. citizen in 1940 (Fig. 4.19).

Notoriously, Einstein had serious problems with several developments in quantum mechanics, starting with Bohr's model of the hydrogen atom in 1913 (although he soon admitted this), in 1925 with the matrix formulation of Werner Heisemberg, and then with Max Born's probabilistic interpretation. In a 1926 letter to Born, he wrote the famous sentence: "*At any rate, I am convinced that God does not throw dice.*" The so-called Bohr–Einstein debates were very important for the understanding of the philosophy of quantum mechanics. In 1935 Einstein had returned to work on this subject, addressing the problem of its completeness as a scientific theory. With Boris Podolsky and Nathan Rosen they published an article in The Physical Review (now known by its initials EPR [59]) in which they proposed a "*Gedankenexperiment*", with two particles that interact in such a way that their properties are strongly correlated. No matter how far apart they later

Fig. 4.19 King Alfonso XIII of Spain (center, with eyes shut) standing next to Albert Einstein. ABC Artistic Collection

separate, the precise measurement of a certain property in one of them should result in an equally precise knowledge of the same property in the other particle, even without the need to disturb the second particle at all. Within the framework of local realism, advocated by Einstein, only two possibilities could fit this situation: (i) the second particle already had that property determined in advance, or (ii) it was the process of measuring the first particle that instantly affected the reality of the property of the second particle. Einstein completely rejected this second possibility (popularly called "action at a distance") [60]. His total belief in local realism led him to affirm that, although the correctness of quantum mechanics was not in doubt, since it was revealed in many experiments, it must necessarily be an incomplete theory. As a physical principle, local realism was shown to be incorrect when, in 1982, an experiment by Alain Aspect confirmed John Bell's theorem, formulated in 1964. The results of these and subsequent experiments have demonstrated that quantum physics cannot be represented with classical versions of the phenomena it describes [61]. This reveals another amazing world, which unfortunately I cannot enter here. Suffice it to say that, although Einstein was wrong about local realism, his very clear exposition of the unusual properties of entangled quantum states has resulted in the EPR paper being among the top ten articles ever published in The Physical Review. It is now considered a pioneering and essential element in the development of the all-important theory of quantum information [62] (Figs. 4.20 and 4.21).

On the eve of World War II, Einstein signed a letter to President Roosevelt alerting him to the enormous danger that "*extremely powerful bombs of a new*

Fig. 4.20 Albert Einstein and Hendrik Lorentz, photo-graphed by Paul Ehrenfest (1880–1933) in front of his home in Leiden, in 1921. Museum Boerhaave, Leiden. Public Domain

Fig. 4.21 Niels Bohr and Albert Einstein. The pictures were taken at Ehrenfest's home in Leiden, the occasion was most likely the 50th anniversary of Hendrik Lorentz' doctorate (December 11, 1925)

type" could pose in Germany's hands, and recommending that the United States prioritize research on such bombs. These finally materialized in the Manhattan Project. Although Einstein always supported the allies, after the war he firmly denounced the idea of using nuclear fission as a weapon. Together with the British mathematician and philosopher Bertrand Russell, he signed the so-called Russell-Einstein manifesto, which highlighted the enormous danger of nuclear weapons.

After his fundamental research on general relativity, Einstein began what would be a long series of attempts to generalize his geometric theory of gravitation to include electromagnetism as another aspect of a single entity. In 1950, he described his "unified field theory" in a Scientific American article entitled "*On the General Theory of Gravitation.*" [63]. Although he continued to be praised for his work, Einstein became increasingly isolated in his investigations. All his efforts in search of a unification of the fundamental forces failed and, what is worse, Einstein ignored the important developments in physics that were taking place at that time. There were many of them, especially regarding quantum fields, elementary particles, and the weak and strong nuclear forces. In their turn, other physicists ignored Einstein's works. They went down another, very different path, which proved much more fruitful in unifying the new interactions that appeared. Einstein's dream of unifying the other interactions of physics with gravity has remained an enormous challenge: it has still not been possible to include gravity in any unification. Until his death on April 18, 1955, Einstein was affiliated with the Institute for Advanced Study at Princeton in New Jersey (Fig. 4.22).

He published more than 300 scientific articles and more than 150 nonscientific works. His intellectual achievements and originality have made the word "Einstein" synonymous with "genius." [64]. In the pre-WWII period, The New Yorker published a cartoon in its *The Talk of the Town* section, saying he was so well known in the United States that they continually stopped him on the street to ask him to explain "*that theory.*" Tired of not being able to take even two steps, he finally figured out how to solve his problem. He decided to say to those who tried to speak to him: "*Sorry, sorry! They always confuse me with Professor Einstein.*" Eugene Wigner compared him to his contemporary colleagues, a select group of the most illustrious mathematicians and physicists, who built the incomparable edifice of theoretical physics during the past century. Wigner wrote, in conclusion, that: "*Einstein's understanding was even deeper than Jancsi von Neumann's. His mind was more penetrating and more original.*" And Paul Dirac viewed the theory of general relativity as "*probably the greatest scientific discovery of all time.*" Why this may be so we will soon realize.

Fig. 4.22 Albert Einstein in his later years (probably 1950s). *Source* Photograph from the Library of Congress Photographs and Prints Division: http://hdl.loc.gov/loc.pnp/cph.3c06042. Author: John D. Schiff. Permission: Copyright John D. Schiff, New York

4.2   The Fundamental Principles of the General Theory of Relativity

After this introduction—which I could not have made any shorter, given the magnitude of such a genius—and in order to understand in what precise way GR could lay the foundations of all modern cosmology, we must now devote some time to trying to understand the essence, or at least the general sense, of its basic principles. This is not an easy task, but perhaps not so extremely complicated, either. The reader should note that, while Richard Feynman claimed, "*no one is capable of understanding quantum mechanics*," he never said the same thing about the theory of relativity.

Einstein built his beautiful theory starting from pure logic. From unprovable basic assumptions, apparently obvious and which he took as the axioms of the new theory, such as the equality of inertial mass and gravitational mass, and the constancy of the speed of light in vacuum—verified in the famous experiment made in 1881 by Albert A. Michelson and then more precisely by Michelson and Edward W. Morley in 1887—and relying on Bernhard Riemann's geometry. And this theory has been proven true in all the experiments carried out to date in our Solar System, in our Galaxy, and even in more distant regions of our Universe. To repeat, GR is a theory that arises

a   b   c   d

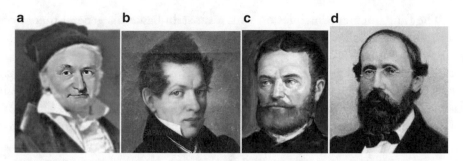

Fig. 4.23 **a** Portrait of Carl Friedrich Gauß (1777–1855) by Christian Albrecht Jensen. Created: 1840. **b** Portrait of Nikolai Lobachevsky (1792–1856) by Lev Kryukov. **c** Portrait of János Bolyai (1802–1860) by Ferenc Márkos. **d** Photo of Bernhard Riemann (1826–1866) taken in 1863. Public Domain

from the purest logic, applied to principles that are naturally taken as true, even without being demonstrable (that is, as axioms) (Fig. 4.23).

I would go even further. The logic and extreme simplicity of the fundamental principles is so overwhelming (they fall under their own weight, as typical Spanish country folk would put it) that it is hard to understand how a theory of such extraordinary richness and complexity can be obtained from such obvious postulates. Note, however, that this point is shared with quantum physics. The secret is always hidden in the magic of differential equations, simple in appearance, but rather complicated to solve, and which admit an extraordinary variety of solutions. Perhaps it reaches its crudest and most devilish expression in the (Navier–Stokes) equations of fluid mechanics. But that is another matter. Let me get back on track.

It was certainly the developments in geometry in the nineteenth century, in particular non-Euclidean geometries, by Friedrich Gauss (1777–1855), Nikolái Lobachevski (1792–1856), János Bolyai (1802–1860), and Bernhard Riemann (1826–1866), which allowed Einstein—with the invaluable collaboration, according to some historians, of his mathematical companion Marcel Grossmann and his wife Mileva Maric (1875–1948)—to formulate his theories of relativity, first the special and then the general one. But we should not forget the important contributions of other great scientists, such as Henri Poincaré (1854–1912), David Hilbert (1862–1943), and Hendrik Lorentz (1853–1928), to name just three of them. Well, it turns out that all modern cosmology is based on Einstein's field equations, which he formulated in 1915, corresponding to his general theory of relativity, and this is what leads many authors to locate the origin of modern cosmology (or at least *theoretical* cosmology) in February 1917, when Einstein first used his equations to build a model of our Universe.

The fundamental principles on which Einstein based his general theory of relativity are as follows:

1. **The equivalence principle**. This can be formulated in various ways. The most canonical version, although framed in a somewhat technical language, is the one that states that, pointwise in space–time, a system with a gravitational field is indistinguishable from an accelerated non-inertial reference system. Otherwise, any instantaneous event of a point-wise nature within a gravitational field can be described by an accelerated observer located at that point, as free movement. That is, there is a certain accelerated observer who has no way of distinguishing whether or not the particles are moving within a gravitational field. Let us take an example: if two parachutists jump, one after the other from a high-altitude plane, in the minutes that elapse while they have not yet opened their parachutes, the second will see the first always in front of him and at the same distance, as if he is not moving at all, as if no gravity is acting on him. Let us retain the image for these minutes: it would be exactly as if they both were astronauts, quietly taking a spacewalk, one below the other, in the absence of appreciable gravity. Einstein came up with this principle after having what he described as "*the happiest idea of all my life.*" Einstein himself explained that it occurred to him in 1907, while working at the Patent Office in Bern. He was sitting in his usual chair when he was suddenly startled by the thought of what would happen to him if he were, at that very moment, falling upright from the roof of his house. He reasoned that, at such an instant, as long as he fell, there would be no gravitational field for him as an observer, at least in his environment. If he held a coin in his hand and released it, it would not fall at his feet! The coin would continue by his hand without detaching itself from it: it would therefore not experience any gravity, according to his perception. Subsequently, the example of an elevator in free fall with a person inside it (accidents like this have actually happened) has been more often used as an alternative to illustrate the same idea.

   Nevertheless, there is yet another way to present the principle of equivalence, which I have used many times in my outreach conferences and popular talks. It has to do with the concept of mass. I remember when Dr. Pedro Pascual arrived as a new full professor in Barcelona and insisted on teaching the first-year degree students. One of his most important lessons was the following: he said that trying to define what the mass of a body is remained one of the most difficult problems in physics (and still is today, despite the fact that we have discovered the Higgs field). But this question is of great importance for cosmologists, for we are continually talking about the masses of stars and galaxies. I will not address the problem in all its difficulty here, but there is a formulation of the equivalence principle which clearly illustrates one of its many facets.

Let us look at two different ways to define the mass of an object. Both come from Newton's laws. On the one hand, we have the universal law of gravitation. It states that two bodies attract each other with a force proportional to the product of their respective masses, M and m, and inversely proportional to the square of the distance that separates them, r. That is, $F = G\,M\,m\,/\,r^2$, where G is the universal gravitational constant (or Newton's constant). This formula (with permission from Einstein's) is without a doubt universally famous, perhaps the most important in all of human history. If an intelligent extraterrestrial civilization ever visited us, this would surely be a key which we could use to try to communicate with it. Let us now consider the case in which M is the mass of the Earth and m that of an object, such as a table (but any object would do). Here m is the mass of the table, relative to the gravitational pull on it from the Earth (its weight, as we call it commonly). Now, there is another equally famous Newtonian formula which tells us that, when we exert a force F on an object, we give it an acceleration a, given by the expression: $a = F/m$. In other words, the more mass the object has, the less acceleration we can impart to it by exerting the same force. Here the mass is what opposes our effort to accelerate the object with our force. The big question is this: is the (gravitational) mass 'm_g' that enters into the gravitational attraction of the object the same as the (inertial) mass 'm_i' that opposes our effort to accelerate it with our own action, or with some other mechanical force? There is no answer to that key question. Of course, over the centuries many experiments have been carried out to check whether or not the two concepts of mass are the same. The first approximation to the answer is "yes," but the margin of error is still important, especially when compared to other physics experiments that reach accuracies of up to eighteen decimal places or more.

From a fundamental viewpoint, there is no reason in the theory to tell us that they must be different. And so, Einstein took this equality of gravitational mass and inertial mass as a postulate of his theory of relativity. In fact, under this form of equality of the two definitions of mass, the equivalence principle dates back to Galileo and it can be said to have originated in his famous Leaning Tower experiment, in which he showed that the acceleration due to gravity did not depend on the shape, color, internal composition, or mass of the bodies he dropped. Although he recognized that the air prevented him from demonstrating his statement more clearly. He also carried out other experiments, with balls and inclined planes with different inclinations. Kepler and especially Newton went further by comparing the gravitational force with others *"produced by animals or other equivalent."* And it is clear that, in his equations, Newton implicitly adopts this principle, by not distinguishing in any way between the inertial mass and the gravitational mass. With his law of universal gravitation, Newton clearly established, confirming what Galileo had said, that the force of gravity does not depend on the characteristics of bodies. Not on any of them, except for one: its mass. But Einstein went further, when he eliminated even the mass: by free falling from the roof of his house (*Gedankenexperiment*) and observing that the mass then just disappeared, as if by magic (the coin no longer weighed on his hand), and with it the gravitational force altogether. Mass had been converted suddenly into space-time *geometry*, by a coordinate transformation, equivalent to speeding up the frame of reference (tied to himself as he fell) in exactly the same way that gravity did. Thus, Einstein demonstrated that the gravity force is not like the other forces mentioned: it becomes a *purely mathematical* property of space-time. More specifically, the mass is associated with its geometric curvature. An exceptional, magnificent, extraordinary discovery, which explains why Einstein said that it was without a doubt *"the happiest idea of all my life."*[1]

Thinking about this for a few minutes, the reader will soon realize that this last formulation is included within the first formulation of the equivalence principle. From a more mathematical point of view, we can express it this way. The space-time manifold can be transformed locally into a Minkowskian one (flat, without curvature, without gravity). Moreover, in a quantum view of things, the metric tensor, $g_{\mu\nu}$, corresponds to the physical field of the graviton. I cannot go further along these paths, no matter how important they are, because they take us away from the road that we wished to follow and which gives sense to this book (although I believe it useful to try to arouse a healthy curiosity in the reader, who may then seek to understand these things on its own).

---

[1] The late Stephen Hawking himself — who made the wonderful discovery that a black hole radiates in accordance with a perfect black body law, and at a temperature involving the most beautiful fundamental constants of nature — wanted to experience, before dying, what Einstein had said. He personally verified, on April 26, 2016, inside an airplane in free fall, with his battered body free at last, like the stars, that the equivalence principle of general relativity was correct: gravity is nothing more than an emergent property of space–time, of the reference system of the Universe.

2. **The speed of light c is constant**. This postulate was inherited from the special theory of relativity, which had taken as its starting point the famous experiment performed by Albert Michelson and Edward Morley in 1887, in which they proved that the ether did not exist. Using the Michelson interferometer, they demonstrated in the laboratory that the speed of light was exactly the same, whether the Earth was approaching or moving away from the emitting source. In other words, the classical kinematic formula for adding speeds (e.g., say of trains running on parallel tracks at different speeds, now in the same direction, then in the opposite), did not apply, when one of the 'trains' was the light. To explain this fact, Hendrik Lorentz introduced the transformation that bears his name and that leads to the famous longitudinal contraction and temporal expansion, coming from a modified law of addition of velocities. And those were the basis for Einstein's construction of his special theory of relativity. This fact has been checked experimentally with extraordinary precision and under various conditions. To the extent that, in 1983, the value of the speed of light in vacuum was officially included in the International System of Units as a constant, and the meter thereafter became a unit derived from this universal constant. Even after such accurate demonstrations, and as in the previous point, there is once again no fundamental reason that explains, a priori, why c is constant, so this fact must be taken as a postulate.

3. **The principle of covariance or general relativity**. The claim is now that the laws of physics are universal, that is, that they are the same at every point in space–time and in any reference system. This is a most natural claim, if we hope to speak of fundamental laws that are valid for the whole Universe. It is an extension of the principle of special relativity which states that the laws of physics take the same form in all inertial reference systems, related by a Lorentz coordinate transformation in space–time.

   In its turn, Einstein's special relativity was the extension to space-time of Galilean invariance or Galilean relativity. This was still a purely kinematic theory, but it already incorporated (a very important distinction) the finiteness of the speed of light, c. It was the very principle of covariance what motivated Einstein to later extend his theory of special relativity. From the physical viewpoint, the principle of covariance implies that there is no procedure to distinguish between various coordinate reference systems. Following his happy idea, which led him to the equivalence principle, Einstein continued down that path until he was convinced that it was possible to build a theory in which the equations were general enough to have the same form in any coordinate system, even when forces intervened (e.g., accelerations).[2] From a mathematical viewpoint, the covariance principle requires that the laws of physics be expressed in a tensorial way, so that the magnitudes measured by different observers can be related by coordinate transformations.

In fact, a first formulation of the principle of relativity dates back to Galileo, who stated his principle of invariance according to which the laws of physics are the same in all inertial reference systems.[3] He illustrated it with the wonderful example of a ship that sails placidly without changing its speed or course (quiet, please, it is Galileo himself who speaks):

> Lock yourself up with a friend in the main cabin, under the deck of a rather large ship; and take with you flies, butterflies and other small flying animals. Hang a bottle so that it empties, drop by drop, into a large container underneath. Make the boat go at the speed you prefer, but always the same: a perfectly uniform movement, without fluctuations in one direction or the other. The drops will fall into the said container, without deviating at all, although the boat has advanced while the drops were still in the air. Butterflies and the other flies will continue their usual flight from side to side, as if they never tired of following the course of the ship, however fast it goes; and it will never happen that they focus on the stern of it.

4. **Zero-torsion hypothesis ($\nabla_X Y - \nabla_Y X = [X, Y]$).** We have already stated the three fundamental postulates of the theory. From there, and in order to simplify it to the maximum while still preserving the previous principles, Einstein introduced some more criteria of a purely mathematical nature and mainly guided by the Occam's razor principle. I will not dwell on them in any detail. The first is to impose that the connection torsion is zero, or that the corresponding Christoffel symbols are symmetrical. This condition can be relaxed, and it has sometimes been, in fact, giving rise to new theories, such as the one called Einstein-Cartan theory. And it is also the case in more ambitious theories in which, in principle, it is even possible to quantize gravity, as in string theories.

5. **Reduction to Newton's laws.** Finally, a very important, and also natural, criterion that the final equations had to satisfy was that they respect all previous physics without any change, in this case, classical mechanics, i.e., Newton's laws. It is an extremely important consideration, even of a philosophical nature, that the fundamental revolutions of physics (or

---

[2] And not only in those that move in a straight line and at a constant speed (Galilean case). Accelerated and arbitrary changes of direction (real and inertial or dummy forces) are now possible.

[3] Recall that, in an inertial system, the reference point and the orientation of the axes are arbitrary, but the velocity must be linear and constant.

revolutions of fundamental physics, which comes to the same) never completely eliminate previous theories, well established in their time and domain of applicability. The theory of relativity had its origin in the verification of the fact that the speed of light (and indeed the speed of propagation of all electromagnetic interactions) was not infinite; and that there were no instantaneous interactions "at a distance." For phenomena that occur at speeds much lower than the speed of light, Newtonian physics continues, and will continue forever, to be perfectly valid to a good approximation, good enough for all purposes in this regime. Consequently, all the expressions of relativistic theories must reduce to expressions of Newtonian physics when one takes the limit as c tends to infinity ($c \to \infty$). This condition is essential in order to be able to define (or identify) the constants that appear in the new theory, when comparing them with the already well-known ones of the old theory.

Two Important Observations

First. The first and third principles, of equivalence and general covariance, respectively, are the two basic postulates of the general theory of relativity (apart from the second, equally crucial but already inherited from special relativity). There has been much discussion about whether they are independent or not. The answer is quite tricky. The two principles are always presented as independent, and in fact they are completely so, in their formulation. But it so happens that, in practice, they are actually connected by inaccuracies that come from the approximations made when formulating the equations of general relativity, which, for simplicity as we have said, are of *second order*. As a consequence, the equivalence principle is *approximate* (in its implementation in Einstein's equations) and is only accurate up to second order terms. Gravity differs in a predictable way from one place to another, as a function of the distance from the source that originates it, while other accelerations are usually the same, from one place to another. Instantaneously, accelerations are indistinguishable at one point, but differentials and gradients of higher orders are not identical. And this small error in equivalence introduces problems with covariance, which mean that terms above the second order are truncated. The curvature of space is well represented in the equations, but not the deformations of space–time of higher orders (Fig. 4.24).

It was Einstein himself who first recognized that his final theory was approximate and incomplete. He hoped that other scientists would soon improve it, something that has not happened even today, despite the fact

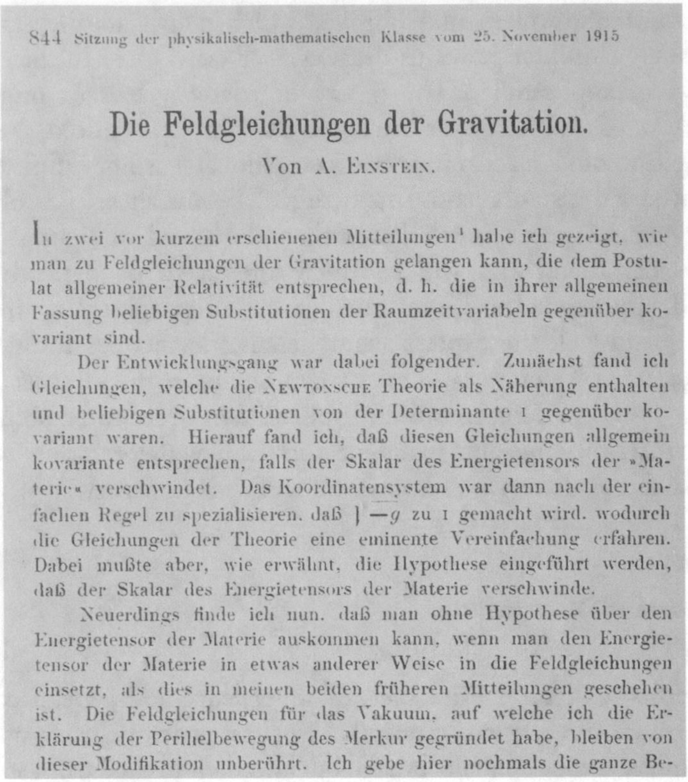

> 844 Sitzung der physikalisch-mathematischen Klasse vom 25. November 1915
>
> ## Die Feldgleichungen der Gravitation.
> ### Von A. EINSTEIN.
>
> In zwei vor kurzem erschienenen Mitteilungen[1] habe ich gezeigt, wie
> man zu Feldgleichungen der Gravitation gelangen kann, die dem Postu-
> lat allgemeiner Relativität entsprechen, d. h. die in ihrer allgemeinen
> Fassung beliebigen Substitutionen der Raumzeitvariabeln gegenüber ko-
> variant sind.
>   Der Entwicklungsgang war dabei folgender. Zunächst fand ich
> Gleichungen, welche die NEWTONSCHE Theorie als Näherung enthalten
> und beliebigen Substitutionen von der Determinante 1 gegenüber ko-
> variant waren. Hierauf fand ich, daß diesen Gleichungen allgemein
> kovariante entsprechen, falls der Skalar des Energietensors der »Ma-
> terie« verschwindet. Das Koordinatensystem war dann nach der ein-
> fachen Regel zu spezialisieren, daß $\sqrt{-g}$ zu 1 gemacht wird, wodurch
> die Gleichungen der Theorie eine eminente Vereinfachung erfahren.
> Dabei mußte aber, wie erwähnt, die Hypothese eingeführt werden,
> daß der Skalar des Energietensors der Materie verschwinde.
>   Neuerdings finde ich nun, daß man ohne Hypothese über den
> Energietensor der Materie auskommen kann, wenn man den Energie-
> tensor der Materie in etwas anderer Weise in die Feldgleichungen
> einsetzt, als dies in meinen beiden früheren Mitteilungen geschehen
> ist. Die Feldgleichungen für das Vakuum, auf welche ich die Er-
> klärung der Perihelbewegung des Merkur gegründet habe, bleiben von
> dieser Modifikation unberührt. Ich gebe hier nochmals die ganze Be-

Fig. 4.24 First page of Einstein's famous work "The Field Equations of Gravitation",
from 1915, in which he establishes the general theory of relativity

that many have tried to do that. General relativity works perfectly well up
to very high energy values, and has recently proved to be very precise in
describing black hole collisions of about 30 solar masses. But if the kinetic
energy is still much higher than that, high enough to fold space–time into
layers, then already problems appear on a theoretical level. A current candidate, among several, to improve the theory (at very, very high energies) is
topological geometrodynamics (TGD). This is a modification of the general
theory of relativity that seeks to solve the problems related to the definition
of inertial and gravitational energies in the hypotheses of general relativity.
TGD is a generalization of superstring models and builds on the previous
work of important scientists other than Einstein, such as Wheeler, Feynman,
Penrose, Josephson, and others. In TGD, physical space–time sections are
visualized as four-dimensional surfaces within an eight-dimensional space.
It is an interesting candidate for the intended generalization of GR and

provides promising results that might resolve the above inaccuracies, although it involves highly complex mathematics.

Second. Einstein's attempt to realize (or materialize) Ernst Mach's ideas in his construction of general relativity was, without a doubt, a very important stimulus for the creation of his theory, even though this does not appear as one of its fundamental principles. It is closely related to the principle of covariance or relativity, although it can be said that it goes even further. In fact, the very name of his theory derives from Einstein's conviction that it was a theory that would finally do justice to Mach's critique of Newton's notion of absolute space; which, in his opinion, should not be such, but should in fact be relativistic or covariant, with respect to the *largest possible transformations* of the space–time coordinates.

Mach's principle, or principle of total relativity, goes even further than the equivalence principle. Total relativity can also be understood as a symmetry principle. It tells us that in the primary equations (in other words, before their solution reveals the crucial influence of distant bodies), we should already put all possible motions on an equal footing, and not only those related by a constant speed or an arbitrary acceleration. It claims that the choice of coordinates is entirely a matter of convention and requires that we remove all intrinsic space–time structures. On that basis, any choice of coordinates should be on an equal footing, since the labels implementing the coordinates could undergo arbitrary variations. But, in general relativity, space–time is not without structure, and it is not true that all coordinate systems are equally good (despite the many statements to the contrary that permeate the literature, beginning with the original article by Einstein). And those are not my own words, but what NP laureate Frank Wilczek states on this point [65]. General relativity always includes a metric field, which tells us how to assign numerical measures to intervals of time and space. It is convenient to choose schemes in which the metric field takes its simplest possible form, because in such frameworks the laws of physics also acquire their simplest form. Posing the problem, Einstein vs Mach, as a matter of symmetry brings it within a circle of ideas that are central to the modern foundation of physics. Indeed, in the standard electroweak model, a Higgs field appears that breaks the local symmetry scale of the primary equations; in quantum chromodynamics, on the other hand, we have a quark-antiquark condensate that breaks both symmetries, along with others; and also in grand unification schemes, generalizations of the idea of symmetry breaking are used (Fig. 4.25).

The symmetry perspective naturally suggests questions that could be fruitful for the future of physics and that go beyond general relativity. It also invites us to contemplate the possibility of primary theories enjoying

Fig. 4.25 Ernst Mach (1838–1916). Heliogravüre by H.F. Jütte, Leipzig. Scanned, image processed and uploaded by Kuebi = Armin Kübelbeck–Zeitschrift für Physikalische Chemie, Band 40, von 1902. Created: 1 January 1903

greater symmetries than those realized in the equivalence principle of general relativity. From this perspective, Mach's principle is the hypothesis that the largest primary theory must include total relativity, that is, the physical equivalence between all the different possible coordinate systems [65]. No one has yet made substantial progress along this path.

Summary
1. Equivalence principle

   - Equivalence of masses: $m_i = m_g$
   - Space-time is a manifold, locally Minkowskian
   - $g_{\mu\nu}$ metric tensor of the graviton field

2. Speed of light is constant, c
3. Principle of covariance or relativity

   - Lorentz invariance, in special relativity
   - On changing inertial frame, equations of physics do not change form

4. Zero torsion hypothesis ($\nabla_X Y - \nabla_Y X = [X, Y]$)

   - Christoffel symbols are symmetric

- Can be relaxed: Einstein-Cartan, string theory
5. Reduction to Newton's laws
   - For compatibility, in order to define the constants.

4.3   Einstein and the Universe in 1917

It has certainly been a fairly long excursion through a vast landscape of fundamental ideas and theories of physics. I have tried to cut a way through, but have only half succeeded. We are now back in 1917, at the very moment when Einstein attempted to apply his field equations of general relativity to the Universe as a whole [66]. This moment can be viewed as the one in which the first stone of the theoretical foundations of modern cosmology was laid. In fact, the whole of cosmology was already there, in those equations, but it took almost a century to unravel them, step by step. Such is the complexity and the richness of the theory.

The reader should bear in mind that, in our time travel, we have slipped back to 1917, the very year in which Slipher's evidence of the high-speed recession of the vast majority of nebulae became conclusive; but we are still three years away from the great debate. As already discussed, in that year of 1917 the Universe was still small (the Milky Way), static (Slipher's results were only known to a few astronomers and nobody had cared to put them to use), and eternal (the first model of the Universe in which it had an origin would only appear 14 years later) (Fig. 4.26).

On February 8 of that year, Einstein published an article in the Journal of Physics-Mathematics Sessions of the Royal Academy of Prussia, in which he applied his general theory of relativity to all matter and energy in space, hence to the Universe as a whole. In the attached images, fragments of some of its pages are shown. He argued in the article that his theory led to the conclusion that the masses in the Universe should bend the space so much that it should already have contracted enormously a long time ago. However, since the Universe was still rather large and such a thing had not happened, Einstein decided to add a term containing a "universal constant," which would act as a kind of 'antigravity' and thus prevent the collapse of the Universe.

But let us read the article directly. Einstein begins by analyzing, on its first page, page 142 of the volume, the case of Newton's gravitation, in which exactly the same problem occurs. He tries to find a solution for it by introducing a constant term, a "*universal constant*" λ (as he calls it), next to the Laplacian of the Poisson equation, showing that this idea will do the

142   Sitzung der physikalisch-mathematischen Klasse vom 8. Februar 1917

Kosmologische Betrachtungen zur allgemeinen Relativitätstheorie.

Von A. Einstein.

Es ist wohlbekannt, daß die Poissonsche Differentialgleichung

$$\Delta\phi = 4\pi K\rho \qquad (1)$$

in Verbindung mit der Bewegungsgleichung des materiellen Punktes die Newtonsche Fernwirkungstheorie noch nicht vollständig ersetzt. Es muß noch die Bedingung hinzutreten, daß im räumlich Unendlichen das Potential ϕ einem festen Grenzwerte zustrebt. Analog verhält es sich bei der Gravitationstheorie der allgemeinen Relativität; auch hier müssen zu den Differentialgleichungen Grenzbedingungen

144   Sitzung der physikalisch-mathematischen Klasse vom 8. Februar 1917

der an sich nicht beansprucht, ernst genommen zu werden: er dient nur dazu, das Folgende besser hervortreten zu lassen. An die Stelle der Poissonschen Gleichung setzen wir

$$\Delta\phi - \lambda\phi = 4\pi K\rho. \qquad (2)$$

wobei λ eine universelle Konstante bedeutet. Ist ρ_0 die (gleichmäßige) Dichte einer Massenverteilung, so ist

$$\phi = -\frac{4\pi K}{\lambda}\rho_0 \qquad (3)$$

eine Lösung der Gleichung (2). Diese Lösung entspräche dem Falle, daß die Materie der Fixsterne gleichmäßig über den Raum verteilt wäre, wobei die Dichte ρ_0 gleich der tatsächlichen mittleren Dichte der Materie des Weltraumes sein möge. Die Lösung entspricht einer unendlichen Ausdehnung des im Mittel gleichmäßig mit Materie erfüllten Raumes. Denkt man sich, ohne an der mittleren Verteilungs-

Fig. 4.26 Extracts from pp. 142 and 144 of Einstein's 1917 publication in which he introduces the cosmological constant, λ

trick. Einstein then transfers it to the case of his field equations for general relativity, which he had deduced in his 1915 paper:

$$R_{\mu\nu} - 1/2\, g_{\mu\nu}R = -8\pi G/c^4 T_{\mu\nu}.$$

And, proceeding analogously, on page 151 he introduces the same type of universal constant, *"eine vorläufig unbekannte universelle Konstante"* (*"a previously unknown universal constant"*), which is now called the *cosmological constant*, in his equations of general relativity. With this, he obtains the new equations:

$$R_{\mu\nu}-1/2\, g_{\mu\nu}R-\lambda g_{\mu\nu} = -8\pi\, G/c^4 T_{\mu\nu}.$$

Next, Einstein argues that the introduction of this additional term into his equations is compatible with all the postulates that we have seen before, in particular with that of covariance, according to which he constructed the theory of GR (in fact, it is the only extra term one could enter). And he reasons, furthermore, that as long as λ is small enough, his new equations will continue to satisfy the fundamental requirement that, when applied to the region of the Solar System, they will still yield results indistinguishable from those of Newtonian physics. All this can be discovered directly by the reader in the image of page 151 of Einstein's work, depicted here. This is a very good lesson to keep in mind: go, whenever possible, to the true origins of history, instead of relying on expressions that have passed from word of mouth, or book to book, through many intermediaries. A second lesson would be: learn foreign languages. On the last page of his work, Einstein explicitly writes that he refuses to compare his model with the results of astronomical observations, limiting himself to purely theoretical considerations. More than one have seen a very serious mistake there, which will be reflected in other similar attitudes throughout his life—and will reappear later in this book (Fig. 4.27).

We must not be fooled, as I have already advanced before, by the apparent simplicity of these equations, which, on the other hand, substantially contributes to their special beauty. As an anecdote, related by Ludwik Silberstein, during one of Arthur Eddington's lectures, he asked him: "*Professor Eddington, you must be one of the only three people in the world who understand general relativity.*" Eddington paused, unable to answer for a moment. Silberstein continued, "*Don't be modest, Mr. Eddington!*" Finally, he replied: "*I am not. Rather, I was trying to guess who the third person might be.*" And, when Einstein made his famous trip to Spain (at least for Spaniards), it is known that only very few scholars were able to understand anything of what he explained, although everyone loudly applauded his presentations.

Actually, this simply looking equation is in fact a set of sixteen coupled partial differential equations, since the subscripts μ and ν each take four values, 0,1,2,3. I cannot go into any more detail, because this would not be the place. Instead, I will now try to convey its deep physical meaning, which revolutionized the science of the day and well into the future.

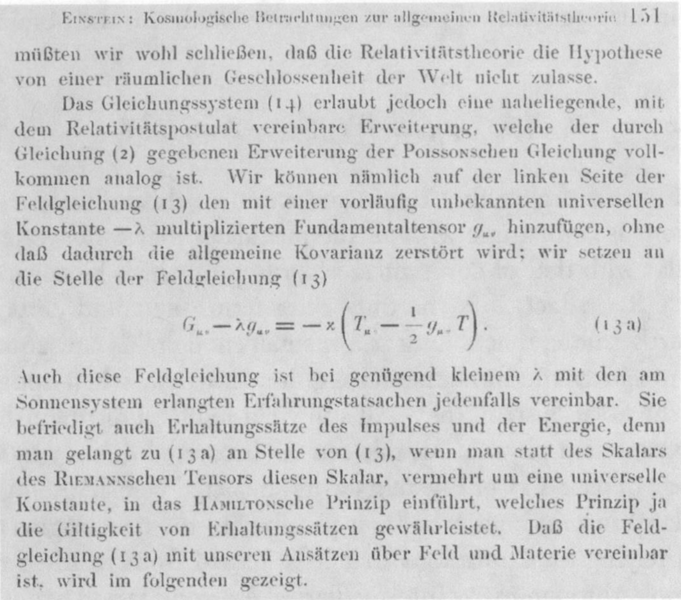

> EINSTEIN: Kosmologische Betrachtungen zur allgemeinen Relativitätstheorie 151
>
> müßten wir wohl schließen, daß die Relativitätstheorie die Hypothese von einer räumlichen Geschlossenheit der Welt nicht zulasse.
>
> Das Gleichungssystem (14) erlaubt jedoch eine naheliegende, mit dem Relativitätspostulat vereinbare Erweiterung, welche der durch Gleichung (2) gegebenen Erweiterung der Poissonschen Gleichung vollkommen analog ist. Wir können nämlich auf der linken Seite der Feldgleichung (13) den mit einer vorläufig unbekannten universellen Konstante —λ multiplizierten Fundamentaltensor $g_{\mu\nu}$ hinzufügen, ohne daß dadurch die allgemeine Kovarianz zerstört wird; wir setzen an die Stelle der Feldgleichung (13)
>
> $$ G_{\mu\nu} - \lambda g_{\mu\nu} = -\varkappa \left(T_{\mu\nu} - \frac{1}{2} g_{\mu\nu} T \right). \qquad (13\,a) $$
>
> Auch diese Feldgleichung ist bei genügend kleinem λ mit den am Sonnensystem erlangten Erfahrungstatsachen jedenfalls vereinbar. Sie befriedigt auch Erhaltungssätze des Impulses und der Energie, denn man gelangt zu (13a) an Stelle von (13), wenn man statt des Skalars des RIEMANNschen Tensors diesen Skalar, vermehrt um eine universelle Konstante, in das HAMILTONsche Prinzip einführt, welches Prinzip ja die Giltigkeit von Erhaltungssätzen gewährleistet. Daß die Feldgleichung (13a) mit unseren Ansätzen über Feld und Materie vereinbar ist, wird im folgenden gezeigt.

Fig. 4.27 Extract from p. 151 of Einstein's 1917 publication "Cosmological considerations on the General Theory of Relativity", in which he introduces the cosmological constant, λ

To do this, I begin with a simpler formula, one that comes from the special theory of relativity, namely:

$$ E = mc^2. $$

It is impressive to realize how an inoffensive expression like this could bring about such an extraordinary conceptual change, whose practical consequences eventually turned the world upside down, thoroughly and forever. The question is, why? As children, we learn in elementary school that it is not possible to compare, add or subtract, amounts that are not homogeneous or commensurable, of the same type. Two apples plus three apples add up to five apples. But what is the result of adding two apples and three pears? Will it be five apple-pears? Not really! And, then in high school, in our science classes, they warn us much more seriously against sums like two kilograms plus three meters per second. And even a popular (Spanish) saying warns us that we should never mix speed with bacon. There is no more serious error in the physics exams than to omit the corresponding units, and carelessly add incompatible quantities. We also learn that two such quantities are mass and energy: it makes no sense to add together grams and ergs. Well, Einstein showed with the preceding formula that, in the area of special relativity, this is indeed feasible, although with the help of a conversion factor, the speed of light squared (a very huge factor, indeed). Sadly enough, atomic and nuclear bombs have demonstrated this formula quite reliably: a few grams of matter are enough to produce and release huge amounts of energy, causing gigantic devastation. The paradigm of classical physics was changed from top to bottom. This does not mean that classical physics does not remain perfectly valid and precise, as it always had been, under the ordinary conditions of any standard laboratory. It is only when we go beyond that, as I have explained before, and enter the new areas that classical physics does not reach, that we need the new physics, to explain the results of the new phenomena that arise there. In fact, it was very difficult to produce the first atomic bomb, e.g., to recreate the conditions needed to manifest this relativistic phenomenon.

Let us now come back down to earth, to a situation in everyday life, to end our example. The keen reader will have had time to find a solution to the problem with apples and pears. When we go to the store, the shopkeeper is perfectly capable of adding the pears and apples that we want to buy. He does it in a very clever way, which has a lot to do with the relativistic equation. Namely, he converts apples into money, applying the conversion factor, which is the price per kg; then he does the same with the pears, adds up the two values, and puts all fruit into the same paper bag (after we pay the total bill, of course). The time I have devoted to what I just explained, in such detail, is not excessive. For, the next step that Einstein took, with his general theory of relativity, was so amazing that we will have to cling to an example as simple as this in order to try to understand it.

Let us turn to the equations of general relativity. We already saw the meaning that Einstein gave to the unknown universal constant λ, as the generator of some kind of 'antigravity', a kind of negative pressure tending to cause an expansion that counteracts the gravitational attraction, to allow for a static universe model. The question may arise whether this term should go to the left or the right of the GR field equations. There is no definitive answer to that; it actually does not matter where we place it, although Einstein originally put it on the left, since λ was a simple constant, without precise physical meaning.[4] I should have started by saying that the terms on the left are all mathematical in nature, in fact geometrical, since they correspond roughly to the curvature of space-time (the curvature tensor $R_{\mu\nu}$ and scalar R, respectively), with $g_{\mu\nu}$ the space-time metric. On the right, the

---

[4] In most current versions of the theory, in which the cosmological constant includes the physical fluctuations of the quantum vacuum state, this term is usually placed on the right. But we are getting ahead of what will be discussed later.

tensor $T_{\mu\nu}$ (which is called the energy–momentum tensor) encompasses all the physical content of the Universe, that is, all the masses and energies of any type which it contains. The universal constants G and c also appear there.

The fact that this is the equation already tells us, without going into greater detail—but using what we have just learned from the previous case—that Einstein now equates pure mathematics, that is, the geometry of space–time (which for Newton was merely the stage, the framework, the immovable coordinate system in which the events and processes of physics took place) with matter/energy (already unified by the special theory of relativity). In other words, the equation tells us that any change in the geometry of space–time influences the mass/energy content itself, and vice versa, a change in the distribution of mass/energy causes a change in the curvature of space–time. From there, we should already be able to understand that we could make a certain amount of mass disappear (on the right), simply by transforming it into an equivalent amount of curvature somewhere in space–time (on the left). And vice versa, the mathematics of space–time, of the coordinate system, its curvature, is also a form of energy. In other words, it has a tangible physical content!

Returning to our example of the grocery: what is the reference system there? Well, it is the person who does the math: the grocer. We may then say that, along with the pears and apples that he has already put inside the bag, according to GR, the grocer himself would also have to get into it! And bye bye shopkeeper! We went to the store in search of healthy energy, and it turned out that, apart from the fruit, the person in charge who did the count, the reference system, that is, the grocer, was also 'energetic,' so we put him in the bag together with everything we bought. I cannot be sure I have made myself clear, but this is the profound philosophy behind general relativity (Fig. 4.28).

Let us now return to Einstein's equations to complete, in a more serious mood, what I have just said. Back again to the first equation, the equivalence of mass and energy. We do not have too much difficulty in accepting this expression, since there are so many laboratory experiments that have confirmed it (apart from the bombs and nuclear power plants). It has been proven quite definitely true: a small mass can be transformed into an enormous amount of energy, and conversely, a large supply of energy actually creates tiny massive particles in high-energy laboratories, such as CERN, in Geneva. In the same way, we should try to find a way to understand the equations of general relativity: the mathematics of space–time. There, a huge curvature, or high-order folding of space–time, or a colossal expansion of the same, can be converted (under certain circumstances) into mass and

Fig. 4.28 Top row: Einstein, Ehrenfest, and De Sitter. Second row: Eddington and Lorentz. Location: W. de Sitter's office in Leiden (The Netherlands). H. van Batenburg–Leiden Archives. Date: 26 September 1923

energy, into a highly energetic plasma of elementary particles or their basic components, namely quarks and gluons. That is precisely what happens in inflationary theories, at the very origin of the Universe (Fig. 4.29).

In accordance with the laws of classical (Newtonian) physics or even those of special relativity, what I have just said is absurd, in fact impossible, since it contravenes the general principle of energy conservation: from 'nothing' it is impossible for 'something' to appear. We could wait a long time at Newton's side contemplating his universal scenery as we sit in his immovable coordinate system, or accompanying Galileo aboard the ship that glides placidly down the Arno river, hoping for a something to appear out of nowhere. At most, an immaterial ghost might rear up out of the shadows, with no tangible body, and certainly no energy. However, if we accompany Einstein in his general theory of relativity, the scenario changes completely: his frame of reference,

Fig. 4.29 Niels Bohr and Albert Einstein. Photo by Paul Ehrenfest (1880–1933), taken at the 1930 Solvay Conference in Brussels

his space–time coordinates, could expand with unusual speed for an unbelievably tiny fraction of a second and then all of a sudden this tremendous expansion could come to an end and originate, in exchange, all the matter and energy of our Universe, which was previously nowhere to be found. In addition, all this could happen while scrupulously respecting the energy balance, that is, the principle of conservation of general relativistic energy, throughout the whole process. This is the whole point. Although this has been just a little taste, a spoiler if you will, of what we will discuss later in much greater detail (Fig. 4.30).

Einstein's general relativity is such an extraordinary theory that all the qualifications we can ascribe to it will fall short. After more than a hundred years, it continues to generate enormous surprises and a considerable number of new Nobel Prize laureates.

Fig. 4.30 Karl Schwarzschild (1873–1916)

4.4   The First Solutions of Einstein's Equations

4.4.1   Schwarzschild's Black Hole Solution

Karl Schwarzschild was born in Frankfurt (Germany) in 1873. He attended a Jewish primary school until he was 11 years old. He was a child prodigy and before turning sixteen, he had already managed to publish two articles, on binary orbits in celestial mechanics [67]. He studied in Strasbourg and Munich, obtaining his doctorate in 1896, for work on the theories of Henri Poincaré. From 1897, he worked as an assistant at the Kuffner Observatory in Vienna. Between 1901 and 1909, he was a professor at the prestigious institute in Göttingen, where he had the opportunity to work with several important figures whom we have already met, such as David Hilbert and Hermann Minkowski. Later he was director of the Göttingen observatory. In 1909, he moved to Potsdam, where he also assumed the position of director of the Astrophysical Observatory, by then the most prestigious position for an astronomer in Germany. In 1912, he was elected to the Prussian Academy of Sciences. At the outbreak of World War I in 1914, Schwarzschild enlisted in the German army, despite being more than 40 years old at that time. In

1915, while serving on the Russian front, he began to suffer from a rare and painful autoimmune skin disease. But nevertheless, while still on the front he managed to write three outstanding works, two on the theory of relativity and one on quantum physics. In his works on relativity he obtained the first exact solutions to Einstein's field equations. A minor modification of his results provided the well-known solution that now bears his name: the Schwarzschild metric, which corresponds to the simplest and most typical black hole. He died on May 11, 1916, very soon after he had sent Einstein his results and established a short postal correspondence with him about them (Fig. 4.31).

Due to its importance, this last point deserves to be elaborated in more detail. It turns out that, while trapped in a hospital near the Russian front, Schwarzschild read Einstein's work, understood it right away, and within a couple of weeks was able to get the first exact solution to the theory. And this while Einstein had only managed to find approximate solutions, and was still working on them. Einstein was very surprised to learn that his

Fig. 4.31 Albert Einstein, in the library of Paul Ehrenfest's house (Leiden, Witte Rozenstraat), where Einstein lived in 1916

field equations admitted exact solutions, something he had hitherto considered practically impossible. Einstein's approximate solution, on which he was
working at the time, gave rise to his famous article, from 1915, on the
advance of Mercury's perihelion. With this aim, Einstein used rectangular
coordinates in order to approximate the gravitational field in the environment
of a spherically symmetric, non-rotating, and uncharged mass. Schwarzschild,
by contrast, chose the simplest possible coordinate system, namely, spherical
coordinates, and was able to obtain an exact solution. Albert Einstein had
submitted his field equations for general relativity on November 25, 1915, at
the weekly Academy meeting, at which Karl Schwarzschild had been present.
Schwarzschild's letter to Einstein is dated December 22 of the same year, less
than one month later! And he wrote it while he was already stationed on the
Russian front! The letter [68] concludes by saying:

> As you can see, the war has treated me with kindness, despite the intensity of
> the shooting; because it has allowed me to get away from everything and take
> this walk through your land of beautiful ideas.

In 1916, Einstein replied to Schwarzschild [69], about his result:

> I have read your letter with the greatest interest. I did not expect that an exact
> solution to that problem could be formulated in such a simple way. I really
> liked your mathematical treatment of the subject. Next Thursday I will present
> your work to the Academy, simply adding a few words of explanation.

Karl Schwarzschild died six months later. His first (spherically symmetric)
solution did not contain any coordinate singularities on the surface that
now bears his name. However, in a second article, he did obtain what is
now known as the "*Schwarzschild internal solution*" (in German, "*innere
Schwarzschild-Lösung*"), valid within a sphere of molecules homogeneously
and isotropically distributed within a spherical surface of radius $r = R$. This
is a case applicable to solids, incompressible fluids, the Sun, and the stars,
viewed as an almost isotropic heated gas; as well as to any gas distributed
homogeneously and isotropically. The solution models gravity outside a star
or a similar large spherical mass, and in particular, around the Sun or the
Earth, so this is a particularly basic and important application of the new
theory of gravity (Fig. 4.32).

Let us now see in more detail how Schwarzschild proceeded to find his
solutions, valid for the Solar System. To do this, he made a few simplifying
assumptions. Even if there are planets, moons, asteroids, and other bodies,

a   b

Fig. 4.32 **a** Reinhard Genzel receiving his Nobel Prize medal and diploma at the Bavarian State Chancellery in Munich. ID: 110,720. © Nobel Prize Outreach. Photo: Bernhard Ludewig, 2020. With permission. **b** Andrea Ghez receiving her Nobel Prize diploma. ID: 111,369. © Nobel Prize Outreach. Photo: Annette Buhl, 2020. With permission

the Sun is by far the largest contributor to gravity in our Solar System. Therefore, Schwarzschild's first simplification was to assume that the Sun was the only source of mass and energy. In fact, to find his first solution, he even ignored the Sun and simply assumed that its mass was negligible. This type of space–time is called the empty one and the corresponding solution the vacuum solution. The equations in this case are greatly simplified: $R_{ij} = 0$, $R = 0$. This space is said to be Ricci flat, which does not mean that there is no curvature, but only that the curvature is particularly simple. Then, for his second solution, Schwarzschild took into account the mass of the Sun, M. Furthermore, he assumed that it was perfectly spherical, which is in fact true up to the fifth decimal place. This again simplifies the analysis enormously. In other words, space–time is spherically symmetric and is the same in all directions of space. Finally, he assumed that nothing in space–time was essentially changing, in other words, that space–time was static. With this, the 16 coupled differential equations of GR were reduced to just one, and a quite simple one at that. Solving this equation, he then found an exact solution for the space–time metric. Later it was proven that it is not really necessary to assume that the space–time is static.

I give below the final solution, as it is always presented today. In spherical coordinates for the space–time, the Schwarzschild metric reads as follows:

$$ds^2 = -c^2 d\tau^2 = -\left(1 - \frac{r_s}{r}\right)c^2 dt^2 + \frac{dr^2}{1 - \frac{r_s}{r}}$$
$$+ r^2\left(d\theta^2 + \sin^2\theta \, d\varphi^2\right); \quad r_s = \frac{2GM}{c^2}.$$

Schwarzschild's space–time is the only spherically symmetric empty space–time. It is important to note that, at large distances (that is, when $r \rightarrow \infty$), the term r_s/r disappears and this metric reduces exactly to the metric of special relativity, Minkowski's flat space–time, written in spherical coordinates. This is because, at great distances, the Sun no longer has any influence and, asymptotically, we obtain the simplest possible solution of the Einstein equations: the vacuum solution, without any mass.

At the other extreme, when r is very small, the two metrics differ enormously. Very close to the star, gravity greatly compresses space–time. The Schwarzschild metric soon became a very useful tool for modeling the behavior of our Solar System. The precession of the orbits of the planets[5] in their movement around the Sun had been known for a long time. It is due in part to the existence of the other planets, whose influence on each other, although only slight, is superimposed on the dominant influence of the Sun. Using Newtonian physics, it had been already possible to explain the precession of all the planets to a good approximation. But, in the early 1900s, it was noted that the value for Mercury's orbit did not actually agree with the calculations. Because it deviated from what was obtained using Newton's theory of gravity, it was thought for a time that there must be another planet very close to the Sun. This planet, known as Vulcan, would have affected Mercury's orbit. However, the hypothetical planet was never found. Here we have an important example of a practical application of Schwarzschild's solution: using it for Mercury, its orbit was easily calculated, and it was found that general relativity corrected Newtonian gravity in the appropriate way, producing a larger precession of Mercury's elliptical orbit, in wonderful agreement with observation. In fact, it was not Schwarzschild but Einstein himself who actually calculated that effect for the first time ever, albeit in a much less elegant way, making approximations to his equations from the outset, as already mentioned.

Indeed, in honor of the truth, and adopting an historical account point by point, it is important to stress that this precession, which has a value of 43 arcseconds per century, was actually obtained by Albert Einstein already

---

[5] The major axis of the orbit does not remain stable, but instead rotates around the Sun.

in November 1915, using his general relativity equations in an approximate way,[6] and after an intensive two weeks' work. Indeed, to get this number was extremely important for him, since it was precisely this result and no other that convinced Einstein that his theory was correct, four full years before Eddington found that starlight curved as it passed near the Sun, during the 1919 eclipse.

Glancing at the Schwarzschild solution one immediately realizes that it presents various problems: there are denominators that cancel when r is equal to 0 or r_s. One could settle this by saying that this space–time was not originally intended to be a good model for such small radius values. For example, the radius of the Sun is 696,000 km, while the value of r_s is, in correct units, just 3 km. And, since the Sun is far from being a vacuum, Schwarzschild's space–time would not be valid inside the Sun, but only outside it. But what would happen if there were something so compact that all its mass occupied a sphere of radius r_s? Today we know that this would be a black hole, namely a region where gravity is so intense, and space–time bends so much, that not even light can escape from it. However, several decades would pass before the Schwarzschild solution would be interpreted as a black hole in such cases.

Anyway, we can now sit back, relax, and enjoy things again. After several pages of rather hard mathematical language and complicated expressions, we can get back to cosmology in its pure state once again. Actually, the concept of black hole, in a broader sense, is very old; at least two and a half centuries older than the formulation of general relativity. Newton's universal gravitation already makes it quite easy to calculate that, if a star were massive enough, its own light could not escape it. Indeed, like most of his contemporaries, Newton considered that light was made up of tiny particles, and for a star with enough mass, the escape velocity could be greater than the speed of light, something which had already been calculated with some precision. We still have a letter, dating from 1783, that John Michell wrote to Henry Cavendish of the Royal Society, in which he talks for the first time about the concept of a 'dark' or 'black' star, in the following terms:

> If a sphere of the same density as the Sun exceeded its size by 500 times, a body falling from infinite height would have acquired, on reaching its surface, more speed than light; all the light emitted by a body like this would then fall again, due to its own gravity.

A few years later, in 1796, Pierre-Simon Laplace, in his famous work "*Exposition du système du Monde*," came to the same conclusion, although

---

[6] That is, not going through an exact solution of his equations, as Schwarzschild did.

it must be added that it only appeared in the first and second editions of this book, and not in subsequent ones. The reason for this is unknown.

Without wishing to linger on this point, it is convenient to show the approximate escape velocities corresponding to various objects in the Solar System. The table speaks for itself. These velocities are compared in the last column with the escape velocity from Earth, which is taken there as unity.

Table of escape velocities (in km/s)

Object	Mass (kg)	Radius (m)	V_{esc}	Resp Earth
Earth	6.0×10^{24}	3.5×10^6	11.2	1
Sun	2.0×10^{30}	7.0×10^8	617.5	55.1
Moon	7.3×10^{22}	1.7×10^6	2.4	0.21
Mars	6.4×10^{23}	3.4×10^6	5	0.45
Jupiter	1.9×10^{27}	7.1×10^7	59.5	5.31
Solar system			~1000	

From this table it is very easy to deduce Michell's statement. However, it was not until 1958 that David Finkelstein correctly interpreted Schwarzschild's solution, and specifically, Schwarzschild's radius r_s, as corresponding to a region of space from which nothing can escape, not even light, which is the fastest and lightest object. Moreover, it was not until nine years later, during a conference given by John Wheeler in 1967, that these solutions were baptized with the suggestive name of Black Holes. There is an anecdote, according to which, in the course of that conference, in which Wheeler was talking about them and their properties, he asked himself, out loud: "*how could we name these objects, in a clear and simple way ...*" At the back of the room, someone from the audience (whose name has been forgotten) raised his voice and proposed to call them "*black holes*". After three seconds, Wheeler nodded: "*Yes. Let's call them that, black holes.*" And the issue was settled forever (Fig. 4.33).

Black holes are real objects that have since captured the imagination of the human mind more than any other science fiction gadget has ever done. I could go on for pages and pages along this path. However, let us just say that, since they are very hard to detect directly (this was achieved for the first time only six years ago), their existence has to be inferred indirectly in most cases. They generally accrete huge amounts of matter around them and sometimes devour nearby stars. The existence and the main characteristics of the black hole at the center of our galaxy, Sagittarius A*, could be determined from the motions of nearby stars, a discovery for which Reinhard Genzel and

Fig. 4.33 Willem de Sitter (1872–1934). Wide World Photos—Photographic Archive University of Chicago. Created in the 1930s

Andrea Ghez shared half of the 2020 Nobel Prize in Physics.[7] In very energetic processes, when black holes are produced by a hyperpernova explosion or by a collision between neutron stars, an emission of so-called gamma ray bursts (extremely violent gamma ray emissions) sometimes occurs. These are among the most energetic phenomena so far observed in the Universe. And the number of black hole candidates increases with every passing day.

4.4.2 De Sitter's Universe

Willem de Sitter was born in 1872 in Sneek, near Leeuwarden, in the Netherlands [70]. He studied mathematics at the University of Groningen and then joined the astronomical laboratory at the same university. At the age of 25, he went to work at the Cape Observatory in South Africa, where he stayed for two years. Later, in 1908, de Sitter was appointed professor of astronomy at Leiden University. In addition, he was the director of the Leiden Observatory from 1919 until his death.

De Sitter made very important contributions to the field of physical cosmology. His most famous achievement in that regard, and for which his name is mentioned assiduously today, is the formulation of the concepts now known as de Sitter space and the de Sitter universe [71]. They arose from a solution that he obtained—possibly as early as in 1916 but more probably in

---

[7] It is noticeable that Andrea is only the fourth woman to obtain this prize in physics. The other half of the award went to Roger Penrose, for his singularity theorems. He will appear later.

February or March, 1917—for Einstein's general relativity equations corresponding to a universe devoid of matter and with just a positive cosmological constant:

$$R_{\mu\nu}-1/2\ g_{\mu\nu}R-\lambda g_{\mu\nu} = 0.$$

This resulted in an empty universe that was expanding exponentially. The value of the curvature was determined by the cosmological constant, and vice versa: $\lambda = 3/R^2$. The solution resulted in either a 4-dimensional hypersphere within a 5-dimensional Euclidean space, or a $(3+1)$-dimensional hyperboloid within the $(4+1)$-Minkowski space–time.

In the language of general relativity, de Sitter's space is the maximally symmetric vacuum solution of Einstein's field equations with a positive cosmological constant (corresponding, in current language, to a positive vacuum energy density and negative pressure). De Sitter was also the co-author of a famous paper with Albert Einstein, in 1932, about the so-called Einstein-de Sitter universe, in which they discussed the implications of cosmological data (which we will see later) for the curvature of the Universe. He also gained renown, in his time, for his investigation of the motion of Jupiter's moons. De Sitter died in 1934.

We have seen before in great detail how Einstein held firm beliefs about what the Universe should be like, and did so with great tenacity. Naturally, this guided him in the construction of his cosmological model. One such belief was that the Universe should be static, since there was no physical reason why it should not be (quite the contrary). Another was that the metric structure of the Universe should be completely determined by matter; in other words, that its metric field should satisfy what he himself called, in 1918, "the Mach principle," since this reflected the revolutionary ideas of one of those who had most inspired him, the physicist and philosopher Ernst Mach. Einstein had read Mach while he was still a student and proclaimed himself a follower. In 1902, when he lived in Zurich, he regularly met with friends of Mach. Although he later realized, as we have seen in some detail, that Mach's ideas and empiricism were very difficult to implement in his equations. For completeness, I should add that Mach also studied the physics of fluids at speeds higher than the speed of sound, and discovered the existence of what was later known as the Mach cone. He established that the relationship between the speed at which a body moves in a fluid and the speed of sound is a very important physical factor (today known as Mach's number) (Fig. 4.34).

Fig. 4.34 The Russian mathematician and physicist Aleksandr Fridman (also written Alexander Friedmann, 1888–1925)

However, it turns out that de Sitter's vacuum solution of Einstein's field equations with a cosmological term is a clear counterexample to Mach's principle, and for this reason, among others, Einstein tried to find a way to discard it at any price. On the one hand, he argued that de Sitter's solution was not static, and on the other, that it exhibited what we now call an intrinsic singularity, which for him meant that it could not actually be free of matter. However, in the end, Einstein had to admit that de Sitter's solution is in fact totally regular and free of matter, and also that it is a genuine counterexample to Mach's principle. But Einstein still had one last card up his sleeve, an argument that would allow him to discard it for another reason: the solution was, in any case, physically irrelevant to describe our Universe, which clearly contains matter in abundance. Today, de Sitter's solution is extremely important in describing the initial and final stages of our Universe, as we shall see later.

Of great cosmological importance until the early 1930s was the so-called "de Sitter effect," which de Sitter formulated in 1917, namely, spectral changes occurring in de Sitter's solution for distant astronomical objects (B stars and spiral nebulae) that made the light emitted by these objects reach the Earth displaced towards the red. This was a quadratic effect, growing larger at greater distances. This is because de Sitter's universe is expanding, and not only that, but the expansion is accelerating rather quickly, making the effect quadratic with distance. During the 1920s, many astronomers were interested in such an effect, considering the various formulations proposed to explain the redshifts obtained by Slipher and others in other terms than Doppler shifts due to relative speeds. Despite the fact that de Sitter's was a universe without matter, this theory and the corresponding effect were preferred for their simplicity by many astronomers, compared with the solutions we will discuss below, which were obtained by Alexander Friedmann,[8] when seeking to explain the observed spectral redshifts. This is a crucial point if we are to understand why expanding solutions and the expansion of the Universe were rejected by almost all cosmologists, and for such a long time.

Anyway, it is quite certain that, historically, the first serious attempts to relate astronomical observations with the geometry of the Universe arose around Sitter's model and its solutions, although with hindsight all of them were unsuccessful. These vain pursuits lasted until the 1940s, more or less. I shall not continue this description here, coming back to this story in more detail later. It will be one of the central points of this book.

4.4.3 The Friedmann Families of Solutions

Alexander Friedmann was born in 1888 in Saint Petersburg, where he remained for much of his short life [72]. He was the son of a composer and ballet dancer and a pianist. He obtained his bachelor's degree from Saint Petersburg State University in 1910, and later became a professor at the city's Mining Institute. One of his fellow students was Jacob Tamarkin, who would later become a distinguished mathematician at Brown University in the USA. Friedmann fought in World War I as an aviator of the Russian Imperial army. Under the revolutionary regime, he became head of an aircraft factory. In 1922, nine years after completing examinations for a master's degree that had been interrupted by the war, Friedmann finally presented his master's thesis. Entitled *"The hydromechanics of a compressible fluid"*, it had two parts, the first on vortex kinematics and the second on the dynamics of a compressible fluid. But it turns out that, upon his return to Petrograd (which was the new name for St. Petersburg) Friedmann had acquired a keen interest in the mathematics underlying Einstein's general theory of relativity. Although published

---

[8] Most likely, because de Sitter was far more popular and his solution was simpler to grasp.

four years earlier, this theory was still not well known in Russia, largely due to the vicissitudes of the First World War and, subsequently, the bloody Patriotic Revolution (Figs. 4.35 and 4.36).

Friedmann was a friend of Paul Ehrenfest. They had known each other during the five years that the latter had spent in Saint Petersburg. Towards the end of 1920, Friedmann wrote him a letter saying:

> I have been working on the axiomatics of the relativity principle, starting from two propositions: a) uniform movement continues to be so for all observers in uniform movement; b) the speed of light is constant (the same for both a static and a moving observer). In addition, I have managed to obtain formulas, for a universe with a spatial dimension, which are more general than the Lorentz transformations.

The Ehrenfest archive at the Lorentz Institute in Leiden [73], The Netherlands, contains several other letters and manuscripts that Friedmann sent to

Fig. 4.35 Portrait of Warrant Officer A.A. Fridman, teacher of the Kiev military school of pilot-observers. Created: 1 August 1916

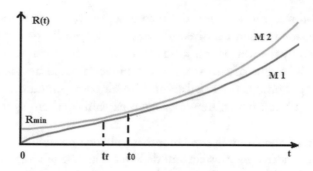

Fig. 4.36 Two possible main scenarios of the evolution of the Universe according to Friedmann (1922). The world M1 shows an expansion from the singularity at t = 0, with an inflection point tf that signifies the existence of two stages in evolution: deceleration and acceleration. The world M2 shows expansion from zero from an initial radius Rmin to infinity. Point t0 corresponds to the current Universe. Source: "The Waters I am Entering No One yet Has Crossed": Alexander Friedman and the Origins of Modern Cosmology, Ari Belenkiy, Fig. 5. Origins of the Expanding Universe: 1912-1932ASP Conference Series, Vol. 471, Michael J. Way and Deidre Hunter, eds. ©2013 Astronomical Society of the Pacific. Reprinted and adapted with permission

Ehrenfest, starting from early in 1922. The translation of a letter he wrote to him in Russian, in April of that year, reads:

I am sending you a brief note on the shape of a possible universe, more general than Einstein's cylindrical and de Sitter's spherical ones. Apart from these two cases, a world also appears whose space has a radius of curvature that varies with time. It seemed to me that such a question might interest you or de Sitter. As soon as I can, I will send you a German translation of this note. And, if you think that the question under consideration is interesting, please be so kind as to endorse it to a scientific journal.

This work, К ВОПРОСУ О ГЕОМЕТРИИ КРИВЫХ ПРОСТРАНСТВ (*On the question of the geometry of a space with curvature*), dated April 15, 1922, does not appear in Friedmann's publication list, which suggests that it was never published. Ehrenfest sent the manuscript—along with an (undated) letter, which Friedmann had written to Hermann Weyl—to the mathematician Jan Schouten, who worked in Delft. Schouten replied to Ehrenfest in a letter dated June 29, 1922, in which he commented critically on Friedmann's analysis (which did not prevent them from collaborating, on another topic of a purely mathematical nature a few years later).

In the same year 1922, and while all this was going on, Friedmann translated his article into German. He had elaborated it further and changed the title to: О КРИВИЗНЕ ПРОСТРАНСТВА (*On the curvature of space*).

He had introduced more clearly the idea of a possible curved and expanding space, and decided to send it directly to *Zeitschrift für Physik* for publication. It was received by the journal on June 29, 1922. In the document, Friedmann showed that the radius of curvature of the universe could be a function of an increasing or periodic time. Friedmann himself commented on the results of that article in a book which he wrote later, explaining them as follows:

> The case of a stationary universe comprises only two possibilities, which were previously considered by Einstein and de Sitter. The case of a variable universe admits, on the contrary, a great number of possible situations. For some of them, the radius of curvature of the universe increases constantly with time. And there are other situations that correspond to a radius of curvature that changes periodically.

Einstein analyzed Friedmann's article rather quickly, as his response to it was received by Zeitschrift für Physik on September 18, 1922:

> The results regarding the non-stationary universe, contained in the work, seem suspicious to me. In fact, it happens that the solution given for this case does not satisfy the field equations.

Friedmann learned of Einstein's critical response from his friend Yuri Krutkov, who was visiting Berlin at the time. And, on December 6, Friedmann wrote Einstein a letter answering his criticism:

> Taking into account that the possible existence of a non-stationary universe is of some interest, I would like to present to you here the calculations that I have made, so that you can verify and critically evaluate them. [Here he details the operations]. If you find that the calculations that I present in this letter are correct, please be so kind as to inform the editors of Zeitschrift für Physik about it. Perhaps in this case you may want to publish a correction to the statement you made, or give me the opportunity to have a portion of this letter published.

However, when Friedmann's letter arrived in Berlin, Einstein had left on a trip to Japan. He did not return to Berlin until March of the following year; although it seems that, for the moment, he did not read (or just ignored) Friedmann's letter. But in May 1923, Krutkov and Einstein met in Leiden, on the occasion of the last master class of Hendrik Lorentz, who was retiring as professor of his own free will. The two met face to face in the house of Ehrenfest, who was going to succeed Lorentz in his chair. There, Krutkov told Einstein the details contained in Friedmann's letter. We know the result of the ensuing scientific discussion through two paragraphs from two letters that Krutkov wrote to his sister in St. Petersburg a few days later. In the first, he says:

> On Monday, May 7, I was with Einstein reading in detail the Zeitschrift für Physik article by Friedmann.

And, finally, in the other letter, dated May 18, 1923:

> I have managed to defeat Einstein in the argument of Friedman's work. Petrograd's honor is saved!

Einstein had admitted his mistake and immediately wrote to Zeitschrift für Physik retracting his previous note:

> In my previous note I criticized Friedmann's work on the curvature of space. However, a letter from Mr. Friedmann, forwarded to me by Mr. Krutkov, has convinced me that my criticism had been based on a mistake made in my own calculations. I now consider that Mr. Friedmann's results are correct and shed new light.

The retraction was received in Zeitschrift für Physik on May 31, 1923. However, this does not mean that Einstein had thereby been convinced that Friedmann's solutions were of any use, that they had something to do with reality (even though they appeared to be mathematically correct). We can say, in fact, that for at least a decade, no one considered Friedmann's work as a possible model for our Universe. It took Einstein himself ten years to finally admit the expansion of the Universe as a physical possibility, despite the astronomical evidence that had accumulated during that time and which I will describe in some detail below.

On the other hand, there is an important fact that is systematically overlooked. In his work, Friedmann obtained various families of solutions and not only "the Friedmann" solution as it is called today, since he did not necessarily impose the restriction that the universe was homogeneous and isotropic. The attached figure (Fig. 4.14) shows just two of his different generic (families of) universes. One of them has an initial singularity and the other begins when the universe already has a finite radius, R_{min}. In addition to these, he found several other cases, including periodic solutions and one with an additional logarithmic term, which implied that the universe had no origin in time, stretching from minus to plus infinity. It was precisely such a solution that

Lemaître first discovered, or rather rediscovered, as we will see later, and he based his model of the universe on it in 1927.

Two years later, in 1924, Friedmann published a second work, also in *Zeitschrift für Physik*: "*Über die Möglichkeit einer Welt mit konstanter negativer Krümmung des Raumes*" (*On the possibility of a world with negative and constant curvature of space*), Zeitschrift für Physik 21, 326–332, 1924. This work completed his previous one of 1922. He thus obtained all possible cases for the values of the curvature of the universe: positive, negative, or null.

Ten years later, Howard Robertson and Arthur Walker proved that, if the universe is homogeneous and isotropic, the only surviving Friedmann family of solutions is the one that originates from a singularity —the curvature may still be positive, negative, or null. After going totally unnoticed for years, this dynamic cosmological model of general relativity would eventually become the *only possible solution for our world*, the standard one, for both the Big Bang theory and the competing steady state theory, as we will see. It was not until the detection of the cosmic microwave background radiation that the steady state theory was ultimately abandoned, in favor of the current Big Bang paradigm.

We may thus say that, in this way, the magnificent framework of general relativity also became the quintessential theory for the description of our universe. To the particular beauty in its conception (which I have previously stressed) was now added the extremely important fact of its being the *only* possible theory (within the framework of the postulates established by Einstein) and having a *unique* solution, the one first found by Friedmann—sometimes called the Friedmann-Lemaître-Robertson-Walker (FLRW) universe. Indeed, only this one is compatible with the cosmological principle, that is, with a homogeneous and isotropic universe.

In June 1925, Friedmann obtained the position of director of the Leningrad Principal Geophysical Observatory. In July, he participated in a balloon flight that set a height record, reaching an elevation of 7400 m [74]. He died soon after, on September 16, 1925, aged 37, from a misdiagnosed typhoid fever. It is said that he contracted the bacteria when he returned from his honeymoon in Crimea, after eating an unwashed pear that he had bought at a railway station [75]. Before leaving Friedmann, we should just mention that the famous physicists George Gamow and Vladimir Fock were among his students.

5

The Hubble-Lemaître Law and the Expansion of the Universe

5.1 Karl Lundmark and Carl Wirtz

In 1924, the same year as Friedmann wrote his second article, the Swedish astronomer Karl Lundmark hypothesized that all spiral nebulae were somehow standard astronomical objects (in particular, he likened them to the Andromeda nebula). This allowed him to deduce, even if only roughly, their intrinsic luminosity from their mass (as Öpik had also done earlier). Then he used the values of a set of such nebulae to determine, from their brightness as seen from the Earth and their mass, the distance from us. He obtained the speeds of the objects from the table drawn up by Vesto Slipher. Lundmark then tried hard to determine a relationship between the speeds and the distances, but had to conclude that, although it was possible that such a relationship existed, he could not affirm this with certainty from the values and the graphic representation he had made. He published an empirical study in which he tried to fit a line; but the dispersion was too high for the result to be reliable and conclusive.

At this point, to complete my description of that period, which immediately precedes the Hubble-Lemaître law, I must mention that Andrzej Wróblewski once called my attention to two important historical documents by Carl Wirtz [76, 77], from 1922 and 1924, which I had overlooked in my previous accounts. In those papers, Wirtz did in fact derive a relationship between the recession of the spiral nebulae and their distance. Wirtz assumed that all the galaxies were approximately the same size and estimated their relative distance from their apparent diameters. For unknown

© The Author(s), under exclusive license to Springer Nature Switzerland AG 2021
E. Elizalde, *The True Story of Modern Cosmology*,
https://doi.org/10.1007/978-3-030-80654-5_5

reasons, Wirtz's papers were disregarded and soon forgotten, although they are certainly mentioned by historians of astronomy, who often refer to the Wirtz spectral shifts (Fig. 5.1).

Wirtz was a careful worker, like Vesto Slipher, who, as we know, had already observed this phenomenon since 1912. In 1918 Wirtz determined a systematic redshift of spiral nebulae which was very difficult to interpret in terms of a cosmological model with a more or less uniform distribution of stars and nebulae (as in a static universe). It was Wirtz who, when analyzing spectral shifts, introduced the K correction (from the German, *Konstante*), attributed for some time to Hubble. This is a corrective factor that must be entered when spectral observations are carried out using a single filter, e.g., the red one. This term is still used routinely in observational cosmology today. But Wirtz's observational evidence that the Universe could be expanding is no longer remembered [78]. About this issue, Wirtz wrote [79]:

> It is remarkable that our fixed star system has such a fast displacement of 820 km/s, and equally strange is the interpretation of the systematic constant k = +656 km. If we attribute a literal interpretation to this redshift value, then

Fig. 5.1 The German astronomer Carl Wirtz (1876–1939) in 1904. Institut für Astronomie der Universität Wien—Austrian Academy of Sciences Press Communications in Asteroseismology, Vienna Observatory, Volume 149, December 2008, p. 127

this would mean that the spiral nebula system is separating at a speed of 656 km/s with respect to the momentary location of our solar system taken as the center.

In 1922, he wrote an article [80] arguing that the results of the observations suggested that the redshifts of distant galaxies become greater than those corresponding to the closest ones, which he interpreted as an increase in their radial speeds with the distance, and that larger masses had smaller redshifts than smaller ones. In 1924, he obtained more accurate results, and interpreted them as a possible confirmation of an increase in radial speeds with distance, but pointed out that this was also compatible with a possible confirmation of the de Sitter universe, the prevailing interpretation by then. Indeed, as I have stressed before, the latter was the 'official' view at the time, according to which the increased redshift was caused by an increase in time dilation in distant parts of the universe and had nothing to do with the speed of the galaxies [81]. Wirtz's conclusion was therefore not clear, since he admitted both interpretations. In any case, in 1936 he wrote a short article alluding to the priority of his conclusion, from 1922, that the radial speeds of galaxies were increasing with their distance from us [82].

5.2   Georges Lemaître

Our next hero in this brief story of the origins of modern cosmology is Georges Lemaître. Lemaître was born in 1894, in Charleroi (Belgium). He soon showed his precocity in mathematics and physics; and his vocation, too, because they say that, before he was ten, he told his parents that he wanted to be a priest. However, his father wisely convinced him that, for the moment, he should concentrate on studying, which he did at the Jesuit school of the Sacred Heart in his hometown. In 1910, the family moved to Brussels, where he learned at another Jesuit school, Saint Michel's College, before entering, in 1911, the Catholic University of Leuven to study engineering. He graduated in 1913 and began working as a mining engineer.

The atrocities he experienced in World War I, in which he served as a volunteer—reaching the rank of sergeant major and being awarded the Cross of War laureate for his bravery—led him finally to become a priest. It is said that once, during the war, he was punished for telling a superior that his ballistic calculations were wrong, and that, at the front, he was sometimes caught reading the Bible or, even more often, articles full of differential equations. After the war, he continued his studies to obtain, in 1920, a doctorate in mathematical sciences, with a thesis entitled "*L'approximation des fonctions de plusieurs variables réelles*" (*Approximation of real functions of several variables*), written under the supervision of the renowned mathematician Charles de la Vallée-Poussin.

That same year he entered the House of Saint Rombaut to become a priest. His teachers, noting his enormous interest in mathematics and physics, suggested that he study the work of Albert Einstein. Lemaître did so, and started learning tensor calculus and general relativity from Arthur Eddington's books. In 1922, he defended a new thesis, "*The Physics of Einstein,*" which earned him a scholarship from the Belgian government to visit the University of Cambridge (England), as a researcher in astronomy. When he was ordained a priest, by Cardinal Mercier, in 1923, instead of practicing as such in a parish or college, Lemaître used the scholarship he had just obtained to travel to Cambridge and learn from Arthur Eddington. He thus became associate researcher in astronomy and spent a year at St. Edmund's House (now St. Edmund's College), at the University of Cambridge. Eddington introduced him to the inner workings of modern cosmology, stellar astronomy, and numerical analysis. More specifically, he suggested that Lemaître do a doctoral thesis applying the equations of general relativity to cosmology. Lemaître obtained two solutions to the problem Eddington had proposed. The first was a reformulation of a solution that Einstein had already given in 1917, for a closed and static universe, with constant mass/energy density; the second had to do with de Sitter's solution, also from 1917, which we have seen before: a universe without mass, dominated by the cosmological constant (the vacuum solution of the theory) (Figs. 5.2, 5.3 and 5.4).

I must add that some sources are quite critical of the relationship between Eddington and Lemaître: not everything was as idyllic as is sometimes claimed. It seems that Eddington paid little attention to Lemaître, once he had given him the problem to be solved. He did not trust his capabilities too much, and in fact the solutions found by Lemaître, which were simple elaborations of others already known previously, did not actually satisfy him.

Perhaps because he felt unappreciated in Cambridge, UK, Lemaître decided to try his luck in Cambridge, Mass. He did this in the same year of 1923, with another scholarship that he had obtained for a research stay at Harvard University, specifically at the university observatory, in Cambridge, Mass. His mentor there was Harlow Shapley, whom we already know well and who had by then gained international renown for his research on spiral nebulae. Simultaneously, he enrolled as a student in the doctoral program, to pursue a new doctorate at MIT (Massachusetts Institute of Technology), which is just two miles away. It is a walk that I myself, as an MIT scholar,

Fig. 5.2 Georges Lemaître around the mid 1930s

November 19, 1925

MASSACHUSETT'S INSTITUTE OF TECHNOLOGY

Cambridge, Mass.

    I)   The gravitational field in a fluid sphere of uniform
invariant density according, to the theory of relativity.

  (2)   Note on de Sitter' Universe.

    3)   Note on the theory of pulsating stars.)

presented by Georges H J E Lemaître

in partial fulfilment of the requirements for the degree of

Doctor of Philosophy in the Department of Physics of the

Massachusetts Institute of Technology

                Approved by

                . . . .

                . . . .

                . . . .

Fig. 5.3 First page of Georges Lemaître's thesis at MIT, Cambridge, MA, USA. Free text (fair use)

— 49 —

UN UNIVERS HOMOGÈNE DE MASSE CONSTANTE ET DE RAYON CROISSANT,
RENDANT COMPTE
DE LA VITESSE RADIALE DES NÉBULEUSES EXTRA-GALACTIQUES

Note de M. l'Abbé G. LEMAÎTRE

1. GÉNÉRALITÉS.

La théorie de la relativité fait prévoir l'existence d'un univers homogène
où non seulement la répartition de la matière est uniforme, mais où
toutes les positions de l'espace sont équivalentes, il n'y a pas de centre de
gravité. Le rayon R de l'espace est constant, l'espace est elliptique de
courbure positive uniforme $1/R^2$, les droites issues d'un même point
repassent à leur point de départ après un parcours égal à πR, le volume
total de l'espace est fini et égal à $\pi^2 R^3$, les droites sont des lignes fermées
parcourant tout l'espace sans rencontrer de frontière (^{1}).

Deux solutions ont été proposées. Celle de DE SITTER ignore la présence
de la matière et suppose sa densité nulle. Elle conduit à certaines
difficultés d'interprétation sur lesquelles nous aurons l'occasion de revenir,
mais son grand intérêt est d'expliquer le fait que les nébuleuses extra-
galactiques semblent nous fuir avec une énorme vitesse, comme une
simple conséquence des propriétés du champ de gravitation, sans supposer
que nous nous trouvons en un point de l'univers doué de propriétés
spéciales.

L'autre solution est celle d'EINSTEIN. Elle tient compte du fait évident
que la densité de la matière n'est pas nulle et elle conduit à une relation

4. EFFET DOPPLER DÛ A LA VARIATION DU RAYON DE L'UNIVERS.

D'après la forme (1) de l'intervalle d'univers, l'équation d'un rayon
lumineux est

$$\sigma_2 - \sigma_1 = \int_{t_1}^{t_2} \frac{dt}{R} \qquad (20)$$

où σ_1 et σ_2 sont les valeurs d'une coordonnée caractérisant la position
dans l'espace. Nous pouvons parler du point σ_2 où nous supposerons
localisé l'observateur et du point σ_1 où se trouve la source de lumière.

Un rayon émis un peu plus tard partira de σ_1 au temps $t_1 + \delta t_1$ et arri-
vera en σ_2 au temps $t_2 + \delta t_2$. Nous aurons donc

$$\frac{\delta t_2}{R_2} - \frac{\delta t_1}{R_1} = 0, \quad \frac{\delta t_2}{\delta t_1} - 1 = \frac{R_2}{R_1} - 1 \qquad (21)$$

— 56 —

Utilisant les 42 nébuleuses figurant dans les listes de Hubble et de
Strömberg (^{1}). et tenant compte de la vitesse propre du soleil (300 Km.
dans la direction $\alpha = 315°$, $\delta = 62°$), on trouve une distance moyenne de
0,95 millions de parsecs et une vitesse radiale de 600 Km./sec, soit
625 Km./sec à 10^6 parsecs (^{2}).

Fig. 5.4 Excerpts from Lemaître's work of 1927, showing what is detailed in the text. Free text (fair use)

have done so many times, to attend seminars at Harvard. During his stay
at Cambridge, Mass., Lemaître traveled a lot around the country, with the
precise objective of meeting, in person, the most important American physicists and astronomers; in particular, according to his biographers, Forest
Moulton, William MacMillan, Vesto Slipher, Edwin Hubble, and Robert

Millikan. Lemaître was not just a mathematician, with the sole aim of solving differential equations, but actually a true scientist, who sought at all costs to construct a cosmological model according to astronomical observations. No wonder then that he became interested in visiting the most prominent astronomers of the day (Fig. 5.5).

Since his time at Cambridge University, UK, he had had the idea of building a cosmological model, and all his efforts were focused on this goal. So, conversations with Slipher, whom he visited at the Lowell Observatory in Arizona, and with Hubble, whom he went to see at Mt. Wilson, CA, were extremely important to him. Both shared with him their latest results, obtained from the astronomical observations they were carrying out. And they graciously gave him their respective tables of velocities (Doppler shifts) and distances (several obtained using Leavitt's law) of the spiral

ANALYSIS OF RADIAL VELOCITIES OF GLOBULAR CLUSTERS AND NON-GALACTIC NEBULAE[1]

By GUSTAF STRÖMBERG

ABSTRACT

List of radial velocities of globular clusters and non-galactic nebulae.—Table I contains a list of all measured radial velocities of globular clusters and non-galactic nebulae, the great majority of the determinations being by Slipher.

Solar motion.—The sun's motion relative to the two classes of objects studied is given in Table II in rectangular equatorial co-ordinates $(-\xi, -\eta, -\zeta)$ and in polar co-ordinates (A_0, D_0, V_0). The value of the sun's motion relative to the nebulae is 344 km/sec. toward $A_0 = 305°$, $D_0 = +56°$ and, relative to the clusters, 329 km/sec. in the direction $A_0 = 320°$, $D_0 = +65°$.

Curvature of space-time.—On the basis of De Sitter's theory of curvature of space-time, we should expect to find for very distant objects a shift of the spectral lines toward the red end of the spectrum. According to Silberstein, we may expect a shift either toward the red or the violet. The correlation-coefficients between distance and radial velocities give no clear evidence of curvature in either De Sitter's or Silberstein's sense, at least up to the limit of distance studied. The only definite correlation is one between radial velocity and position in the sky, indicating a solar motion of 300 or 400 km/sec. in about the same direction for both classes of objects.

The determination of radial velocities of globular clusters and of non-galactic nebulae is very difficult on account of the faintness of the objects and the absence, in general, of bright lines in their spectra; but through the perseverance of Professor V. M. Slipher, a fairly large number of such velocities has been derived. Two reasons prompted the writer to study these velocities. One was the large solar velocity found from these objects, which, in connection with the asymmetry of stellar motions, indicated that a fundamental reference system could be defined by them. The second reason was the desirability of ascertaining whether the velocities give any evidence of a curvature of space-time.[2] Through the courtesy of Professor Slipher it has been possible to make use of his radial-velocity determinations up to a recent date.

In Table I are collected the data on which the computations are based. Slipher's determinations are given without references; for

Fig. 5.5 Gustaf Strömberg's work, used by Lemaître to extract the table of redshifts, which he later adjusted to his expansion model. Free text (fair use)

nebulae, which have already been mentioned before. On the other hand, the more mathematical work he did for his doctoral thesis at MIT, supervised by Harlow Shapley himself—and that I had the privilege of consulting personally during my various stays as a visiting scientist in this extraordinary institute, second to none[1]—culminated in his obtaining a solution of Einstein's field equations. It corresponds to a universe with mass and in expansion, at a constant rate, but without origin or end, since time stretched from minus to plus infinity. Ultimately, it was found that this was, once more, a solution that had already been found previously by Alexander Friedmann in his now famous paper of 1922. Although this was not precisely "*the Friedmann*" solution, as *many wrongly claim* (chroniclers sometimes copy each other without modesty), but another containing an additional logarithmic term which prevents the universe from having an origin in time. All indications suggest that Lemaître did not know about Friedmann's article at the time (he himself said that explicitly), nor had he been informed of the intense discussion between Friedmann and Einstein (although some of my Russian colleagues allow themselves to harbor a few doubts).

Lemaître returned to Belgium in the summer of 1925. Thanks in part to a recommendation from Eddington, he was appointed part-time associate professor of mathematics at the Catholic University of Leuven. In 1927, he defended his thesis at MIT, entitled "*The gravitational field in a fluid sphere of uniform invariant density, according to the theory of relativity; Note on de Sitter' 'Universe; Note on the theory of pulsating stars*", Lemaître, Georges H.J.E. (Massachusetts Institute of Technology, 1927) [83].

During the two years since his stay in Boston, Lemaître had continued to search for a model of the Universe that would agree with the astronomical results of Slipher and Hubble. And he succeeded in fact, and very brilliantly, in an article that later earned him well deserved international fame. The research that he had carried out in part at Harvard, in part at MIT, and in part in Leuven, was finally written up and published, in the same year 1927, in the little-known Belgian journal *Annales de la Société Scientifique de Bruxelles* (Anales of the Scientific Society of Brussels). Its title is "*Un Univers homogène de masse constante et de rayon croissant rendant compte de la vitesse radiale*

---

[1] Allow me a moment to confess that, for me, the true center of the Universe is there. In its cozy inner courtyards, its vast corridors, and its various libraries, where physics and thought, music and algorithms harmoniously intermingle, not to mention the emerging artificial intelligence, and so many other things (including the future, the still to come), I have felt transported to infinity, to another world, like nowhere else I have been to. One morning, while the chalk I was holding in my hand was shaping one of my most successful calculations on the blackboard in my office, oblivious to what had happened at Boston's airport, my wife interrupted me, calling from Barcelona, to give me the news of the attack on the twin towers. Our oldest son defended his doctoral thesis at MIT, with honors, a couple of years later.

des nébuleuses extra-galactiques" (*A homogeneous universe of constant mass and increasing radius that explains the radial velocity of extragalactic nebulae*) [84]. In this work, he presented his truly revolutionary idea, which he derived from general relativity, that the Universe is in fact expanding; and this was in perfect agreement with the astronomical data of Slipher and Hubble. He used the spectral Doppler shifts of the spiral nebulae as evidence of a cosmic expansion, matching redshifts and the distances of 42 nebulae to deduce the value of the expansion rate (today called the Hubble constant). In this work, he not only obtained Hubble's law from the two tables, of speeds and distances, that the two astronomers had provided him with—and with a value of the Hubble constant very close to the value obtained by Hubble two years later, in 1929—but what is more, he correctly interpreted the redshifts as due to the expansion of the Universe. He left behind all the other theoretical physicists and astronomers far behind, with a clear and precise intuition. It was absolutely amazing, almost unthinkable in 1927. For the data were the same as other astronomers had at their disposal, but every one of them persisted in misinterpreting them!

Everything is there, in this precious work of just eleven pages, in French, for any unbeliever who knows this language and wants to check it out. However, in doing so, the reader will immediately spot an apparently serious omission: in Lemaître's 1927 article no mention is made of Vesto Slipher's redshift table (which, incidentally, has contributed to the fact that he has been so neglected). This may seem strange, since Lemaître knew Slipher's results perfectly, from his visit to the Lowell Observatory in Arizona during the period of his thesis at MIT. Instead, in deriving Hubble's law, Lemaître takes the radial velocities from a table due to G. Strömberg [85]. However, if as good historians we dig a little further, it is only necessary to read the first page—even just the first two lines of Strömberg's paper—to realize that the whole thing is once again an extraordinary tribute to Slipher: "*The vast majority of determinations are from Slipher,*" or obtained "*through the perseverance of Professor VM Slipher,*" and so on (Fig. 5.6).

The "Fifth International Solvay Conference on Electrons and Photons," which took place in Brussels in October 1927, became extraordinarily famous over time. It is probably the highest-level scientific conference ever held: 17 of the 29 participants appearing in the much-famous photograph of it had won, or would later win, the Nobel Prize (Madame Curie did so twice). Lemaître was there, and during a break from the sessions, he cornered Einstein to deliver his recently published article and his conclusions. He took the opportunity to tell Einstein face to face that he had discovered a solution to his field equations of relativity which would correspond to an expanding universe;

Fig. 5.6 Participants of the Fifth International Solvay Conference on Electrons and Photons, Brussels, October 1927. Photograph by Benjamin Couprie, Institut International de Physique Solvay, Brussels, Belgium. From back to front and from left to right: Auguste Piccard, Émile Henriot, Paul Ehrenfest, Édouard Herzen, Théophile de Donder, Erwin Schrödinger, Jules-Émile Verschaffelt, Wolfgang Pauli, Werner Heisenberg, Ralph Howard Fowler, Léon Brillouin, Peter Debye, Martin Knudsen, William Lawrence Bragg, Hendrik Anthony Kramers, Paul Dirac, Arthur Compton, Louis de Broglie, Max Born, Niels Bohr, Irving Langmuir, Max Planck, Marie Skłodowska Curie, Hendrik Lorentz, Albert Einstein, Paul Langevin, Charles-Eugène Guye, Charles Thomson Rees Wilson, Owen Willans Richardson. Benjamin Couprie. Created: 1 January 1927, restored

and that it perfectly fit with the latest astronomical observations. He added, furthermore, that he had managed to demonstrate that the static solution of the Universe that Einstein had obtained was, in fact, unstable. Ultimately, the Universe could not be static; it had to be expanding! After examining Lemaître's work, the answer that Einstein gave him was the following: he had not found any error in the mathematical formulas, but the physical interpretation of them, the idea that the Universe was expanding, did not make the slightest sense—it was *abominable*, as Lemaître himself later detailed in French. Looking at the equations, Einstein had recalled that Alexander Friedmann had been the first to suggest such an unrealistic possibility five years earlier. What made things worse now was that he did not want to pay any attention to the clear evidence that Lemaître showed him, based on the results of astronomical observations. Though, let me repeat, these observations still had to be interpreted correctly, which was rarely the case at the time.

At this point, the reader should pause for a moment to reflect seriously on this surprising fact. To Einstein, a genius, the creator of general relativity, the master of space and time, of its contractions and dilations, the discoverer of the possible existence of entities as esoteric as gravitational lenses and gravitational waves, it took him another four years (until 1931) to become convinced that the Universe was indeed expanding. How to understand such an attitude, so much stubbornness? It must seem strange to us now, even incredible. Rather, his attitude clearly illustrates a couple of issues. To begin with, it supports the conviction of the author of this book that the expansion of space was, at that time, an *extremely revolutionary* idea, which no one accepted for years and that many (like Hubble himself) would never understand! Lemaître was for a certain time the only man on Earth who was comfortable with the idea that physical space was in fact expanding. Another question that we could consider now is this: to what extent was he just lucky to find the correct interpretation of the astronomical results, based on Einstein's fundamental theory of gravity? Whatever the situation, it was terribly difficult for him to convince other colleagues of the important discovery he had made.

Almost all historians attribute this, probably copying each other again, to the fact that Lemaître's work was published "in French" and "in an obscure Belgian journal," which no one read. The truth is much more complex and it has many facets. Here are a few compelling thoughts. First, Öpik's work which I mentioned earlier appeared in the most prestigious cosmology journal, *The Astrophysical Journal*, and no one seems to have paid any attention to it, either. Second, Friedmann's work was published in a highly regarded physics journal, the *Zeitschrift für Physik*, and apparently no one paid any attention to that. Third, Einstein, the greatest theoretician of the time, did have precise and detailed knowledge of both Friedmann's and Lemaître's work. He examined both papers in full detail, but he was not convinced by either of them. It can be argued that, in the first case, these were simple mathematical solutions, possibilities that perhaps had nothing to do with reality, with the true world. But note in particular that Lemaître provided him with clear astronomical evidence that perfectly matched the solution of his own equations! How is it that, even so, he was not convinced? And, I repeat that it was not only him, who was not convinced, but none of the other cosmologists of the time to whom Lemaître spoke or sent his paper. Here is my answer once again: the crucial issue was the interpretation of the astronomical results, of the redshifts.

We should say yet again that the idea that the universe expands is weird, as I always point out in my writings and talks, although we now claim that it should be understood this way by students at school. It is even more so than the concept of light deflection by a mass (gravitational lensing effect) or space–time gravitational waves, which are already very strange notions to *understand*, and not just recite, of course, like parrots without understanding a word! Einstein himself, the father of the whole theory, had no room available in his mind for an expansion of the fabric of space-time. And his fellow cosmologists spent all their available time trying to find alternative interpretations for the huge redshifts of Slipher and Wirtz, other than the one, which now seems to us all to be the only one possible! Redshifts were usually interpreted as associated with those appearing in de Sitter's solution (much more famous than those of Friedmann or Lemaître), and which had nothing to do with the speed of the objects. Throughout his life, Hubble believed (as we will see in more detail later) that these speeds were undoubtedly "fictional" or "apparent," not real, and he always described them this way in his work. Another great astronomer, Fritz Zwicky, who will later appear as the first discoverer of enigmatic dark matter[2] argued that photons simply lost energy as they passed through the universe until they reached us. They gradually yielded their energy to the intergalactic medium through which they traveled, through a process analogous to Compton scattering, leading to a progressive reddening of light (a perfectly plausible physical phenomenon), so that the further away the nebula that emitted them, the more the light that reached us deviated towards the red. And this, again, had nothing to do with the speed of the nebula! Others simply suggested various versions of the reddening of light with distance—collectively, it was called the "tired light" hypothesis—without even claiming to provide a concrete physical explanation. Many astronomers proposed mechanisms to explain redshifts without in any way accepting the expansion of the universe. It was not until the late 1930s that, little by little, they increasingly adopted the view that the redshifts were Doppler shifts and that the universe was expanding (Fig. 5.7).

In short, we have to take all this into account before making simplistic, hasty, and unfounded judgments. Once again, the final moral is the same as in many other situations: the scientists of the time were not stupid at all, just the opposite. It so happens that things, viewed in perspective, through the filter that provides us everything we now know, look completely different from when viewed from within, with only the knowledge they had a hundred years ago. In effect, we must learn how to travel back in time (which is no

---

[2] It was precisely he who gave it this name: *dunkle Materie*.

RADIAL VELOCITIES OF TWENTY-FIVE SPIRAL NEBULÆ.

Nebula.	Vel.	Nebula.	Vel.
N.G.C. 221	− 300 km.	N.G.C. 4526	+ 580 km.
224	− 300	4565	+1100
598	− 260	4594	+1100
1023	+ 300	4649	+1090
1068	+1100	4736	+ 290
2683	+ 400	4826	+ 150
3031	− 30	5005	+ 900
3115	+ 600	5055	+ 450
3379	+ 780	5194	+ 270
3521	+ 730	5236	+ 500
3623	+ 800	5866	+ 650
3627	+ 650	7331	+ 500
4258	+ 500		

Fig. 5.7 Radial velocity table in km/s of 25 spiral nebulae, published by V.M. Slipher in 1917: Nebulae, Proceedings of the American Philosophical Society, vol. 56, p. 403–409, 1917, Table 1. Bibcode: 1917PAPhS.0.56.0.403S. Free text (fair use)

simple matter), to situate ourselves in the epoch in question, forgetting for a moment everything we now know. And it is then, and only then, that we may clearly understand that Lemaître was very lucky to hit the right nail on the head in such a precise way in 1927. And he did it before anyone else; earlier, in particular, than other physicists and astronomers who actually knew a great deal more than he did. Einstein once said to Lemaître, at the end of a lecture given by the latter: *"You have just given the most beautiful explanation of the creation of the Universe that I have never heard."* Although that is another topic, on which I will expand later (Fig. 5.8).

But let us not kid ourselves; during his life, Lemaître was considered by many colleagues as a second-rate cosmologist. Both he and Friedmann himself were, by training and according to the works they wrote, essentially mathematicians, and the results they obtained were mostly mathematical solutions to differential equations. Friedmann limited himself to observing that his solutions could be interesting to describe the world physically, as a serious alternative to the Einstein and de Sitter universes. Lemaître, for his part, was more ambitious. He clearly went in search of a cosmological solution that included the best properties of both worlds and avoided the problems of each[3]: a description of the universe that also matched the astronomical results. And, in the end, without having made original contributions of great value (during his life he sometimes received quite derogatory ratings

---

[3] This is precisely how he describes his intentions in his 1927 work.

TABLE 1

NEBULAE WHOSE DISTANCES HAVE BEEN ESTIMATED FROM STARS INVOLVED OR FROM
MEAN LUMINOSITIES IN A CLUSTER

OBJECT	m_s	r	v	m_t	M_t
S. Mag.	..	0.032	+ 170	1.5	−16.0
L. Mag.	..	0.034	+ 290	0.5	17.2
N. G. C. 6822	..	0.214	− 130	9.0	12.7
598	..	0.263	− 70	7.0	15.1
221	..	0.275	− 185	8.8	13.4
224	..	0.275	− 220	5.0	17.2
5457	17.0	0.45	+ 200	9.9	13.3
4736	17.3	0.5	+ 290	8.4	15.1
5194	17.3	0.5	+ 270	7.4	16.1
4449	17.8	0.63	+ 200	9.5	14.5
4214	18.3	0.8	+ 300	11.3	13.2
3031	18.5	0.9	− 30	8.3	16.4
3627	18.5	0.9	+ 650	9.1	15.7
4826	18.5	0.9	+ 150	9.0	15.7
5236	18.5	0.9	+ 500	10.4	14.4
1068	18.7	1.0	+ 920	9.1	15.9
5055	19.0	1.1	+ 450	9.6	15.6
7331	19.0	1.1	+ 500	10.4	14.8
4258	19.5	1.4	+ 500	8.7	17.0
4151	20.0	1.7	+ 960	12.0	14.2
4382	..	2.0	+ 500	10.0	16.5
4472	..	2.0	+ 850	8.8	17.7
4486	..	2.0	+ 800	9.7	16.8
4649	..	2.0	+1090	9.5	17.0

Fig. 5.8 Table of distances in Mpc of spiral nebulae, published by E. Hubble in 1929: A relation between distance and radial velocity among extra-galactic nebulae, PNAS March 15, 1929 15 (3) 168–173, Table 1. Free text (fair use)

for this), since all the solutions he obtained (both at Cambridge, UK, and at MIT) had been previously discovered (and none of them bears his name today), or having ever used a telescope to make observations of the cosmos (he used the data tables that astronomers graciously provided him with), he learned how to link everything, theory and observation, like the best of scientists, in an absolutely masterful way and anticipating by several years all the great minds of the day. To all this it must be added that Lemaître was always very honest, both with himself and with others: he never claimed priority over Hubble's (empirical) law, which he had discovered two years earlier, nor bragged of the masterful interpretation he had made of it based on the solution he had found for Einstein's equations. That, and no other, is the reality of the facts, putting the dots on the i's and crossing the t's and giving each one his due. For it is so disappointing to read in the literature derogatory stories about Lemaître, just as some others now extol, cloyingly, everything he did.

5.3   Hubble's Law

As I previously stated, Edwin Hubble was, in all probability, the most influential astronomer of his generation. In the late 1920s, comparing the redshift table of 25 spiral nebulae published by Vesto Slipher in 1917 [86] (and which by then had even appeared in Eddington's famous book [87]), and his own distance table for the same nebulae [88], Hubble obtained the law, now so famous, that bears his name, and published it in 1929. In his article he shows a fairly clear linear relationship between distance and redshift, which he interpreted, following Slipher, as velocities. He used Slipher's data, but added some more that Milton Humason had measured at Mount Wilson. He found the distances to the nearest nebulae using Cepheids as standard candles. At greater distances, Hubble used the brightest individual stars that he could resolve for this purpose and assumed that they had the same brightness in all nebulae. And, at even greater distances, he used the luminosities of the nebulae as a whole, something other astronomers did, as we have seen. Hubble's results for the distance-redshift relationship (and not at all the fact that it translated into a radial distance-velocity relationship, as I will discuss later) were quickly accepted by the astronomical community. This relationship is known as "the Hubble law," although, as already discussed, it was anticipated (albeit not so clearly and convincingly) by other astronomers. When the two tables, one of redshifts and one of distances, are placed next to each other, one immediately realizes how easy it is to adjust the values to a straight line, $V = H_0 D$ (that same proportionality is what Lemaître had obtained two years earlier), with H_0 being a constant, which until recently was called the Hubble constant. Hubble did not mention in his article, however, that Slipher was the author of the redshift table; his name does not appear anywhere in the paper. And that point is very important, because this is why, for many years, and even *today*, in many references to his work, it is mistakenly considered that it was Hubble who produced both tables: the distance table and the redshift table. This is how it is put in many books and articles, and I personally even heard a recipient of the Nobel Prize in Physics affirm this falsehood, before a very large audience who may well have believed it without question (Figs. 5.9 and 5.10).

The facts should never be so flagrantly misrepresented, not even for the sake of simplification or lack of time. As John Peacock explains in much detail [89], Hubble was actually fortunate in several ways (which the reader will find in this reference) that he was able to establish his law of proportionality with the material he had at hand. He later admitted that he was in fact pursuing earlier searches (such as Lundmark's and others) for a correlation between

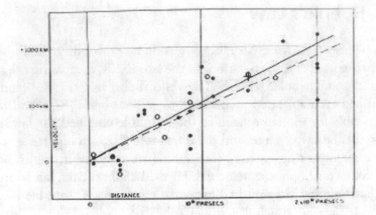

Fig. 5.9 Hubble's law, which Hubble obtained by adjusting both tables, in 1929: Edwin Hubble, A relation between distance and radial velocity among extra-galactic nebulae, PNAS March 15, 1929 15 (3) 168–173, Fig. 1. Free text (fair use)

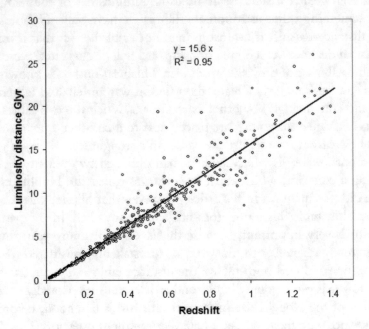

Fig. 5.10 A present day calculation of Hubble's law: Luminosity distance versus redshift of supernovae obtained using the redshift adjusted distance modulus on data from http://supernova.lbl.gov/Union/. *Source* Yheyma—Own work. CC BY-SA 4.0

redshift and distance. Hubble also admitted that these studies were explicitly motivated by the prevailing theory at the time: the de Sitter effect. He had in mind no idea at all about the Universe expanding.

Two years later, in 1931, Hubble and Humason increased the number of redshifts until they reached much higher values, corresponding to speeds almost twenty times higher, of 20,000 km/s [90]. However, their distance measurements were again based on the same (and unwarranted) assumption that Lundmark (and also Öpik, as we saw) had made in 1924, namely that galaxies could be treated as standard objects to calculate their distances from their mass and brightness, in the absence of other, better justified distance estimates. Again, according to Peacock, one could have imagined that the 1931 article would have been greeted with considerable skepticism and criticism for this. But, by then, it had already become apparent to the community that Hubble had indeed demonstrated the linear relationship in his 1929 article: the further away a galaxy was, the faster it moved away from us. And in the minds of astronomers, these results had already shattered Isaac Newton and Albert Einstein's concept of a static Universe. If Hubble was right, the visible objects in the universe were quickly moving away. At least, this is what it seemed (but please go on reading).

5.4   The Interpretation of Hubble's Law

Hubble's law was certainly a remarkable achievement, even if it was only a first step towards the actual scientific theory, since it was a purely empirical result. It said nothing about the physical explanation of the effect itself, about the reason why the extragalactic nebulae would drift away at such tremendous speeds. Even worse, all the given explanations were *wrong*, as we have seen. Moreover, once the spectral shifts are correctly interpreted as Doppler shifts (redshifts are speeds), that is, as real speeds of celestial objects (we have seen before that for several years there was a widespread refusal to accept this interpretation), then the visible universe was no longer static; its volume increased as distant objects recoiled at such high speeds. Now a "subtler" (but crucial!) dichotomy between two new possible interpretations appears, namely: (a) celestial objects are moving away (like somebody who runs away from us, say), or (b) celestial objects have in fact small peculiar motions, but they are embedded in a rapidly moving reference frame (the person is on a high speed train while walking in a leisurely manner down the corridor). Distinguishing between the two explanations was the crucial point in finally turning the empirical law into a scientific theory.

Repeating the concept in a more professional language (the importance of the case requires it), once the interpretation of redshifts as speeds had been established (as Slipher did, from the first day he used his spectrograph in 1912) the key question was now the *interpretation* of Hubble's law:

(a) Do the high escape velocities of the spiral nebulae correspond to actual displacements of the celestial objects?
(b) Do they correspond to the motion of the reference system, in the sense that it is space itself that expands?

Of course, the obvious answer is that, in the end, both things are true since both phenomena contribute to the spectral shift. Even today, this is a very difficult problem to solve in astronomy: separating these two components on maps of the observed spectral Doppler shifts of astronomical objects is technically very complicated. The motion of a given object can be translated into small redshifts or blueshifts, depending on the direction it has. And they can be completely dominant if the galaxy is close to the Milky Way. However, we are talking here about the motions of very distant galaxies where we know that the second interpretation (and the corresponding contribution) overwhelmingly prevails. But this idea of the expansion of the universe was extremely difficult to accept at first, and only Lemaître did so at the beginning, in his 1927 model, where he calculated redshifts directly from his expanding space model (Fig. 5.11).

From recent studies by historians of cosmology, it now seems clear that Hubble never actually believed that the Universe was expanding. There is no doubt that he never said anything about expansion in any of his works. In a letter he wrote to Willem de Sitter, in 1931, he stated his thoughts on speeds by saying [91]:

> ... we use the term 'apparent speeds' to emphasize the empirical character of the correlation. The interpretation, we believe, I must leave to you and the few who are competent to discuss this matter with authority.

A second important observation should be made here. It is stated in many books and articles that it was precisely Hubble who convinced Einstein that the Universe was expanding; and that this happened when the latter visited the Hubble in Mt Wilson in 1931, during his famous tour of the United States that year. However, by examining in detail Einstein's notebook (his famous *Tagebuch*) and other writings from that period, it has been established with certainty that Hubble never spoke to him about the expansion of the universe. On the contrary, it is now clear that Einstein finally reached

Fig. 5.11 Albert Einstein with Edwin Hubble, Walther Mayer, and Walter Adams at the Mount Wilson Observatory, in 1931. Image courtesy of the Observatories of the Carnegie Institution for Science Collection at the Huntington Library, San Marino, California. Digital Collection Photographs, Huntington Digital Library. Unique Digital Identifier 31,317. Fair use

his conviction that the universe was expanding in the course of that same year, through conversations maintained with Eddington, Tolman, and de Sitter [92]. In the end, they managed to convince him, at long last, that his static model was unstable and that it was simply a fact that the Universe was expanding. As we have already seen, Einstein had introduced his famous cosmological constant in 1917, in his attempt to obtain a static Universe model. In 1931 he abruptly renounced this model, removed the cosmological constant, and embraced Friedmann's solution, without the constant. That happened 14 years after he had introduced it.

George Gamow related in *Scientific American*, in 1956, that Einstein had told him many years before that the idea of the cosmic repulsion associated with the cosmological constant had been the greatest blunder of his life (*"Die grösste Eselei meines Lebens,"* in German). For several decades, this was the only existing testimony of such a claim, and some questioned it due to the fertile imagination Gamow displayed on many occasions. But recently, thanks to the work of the tireless historians once again (in this case, Cormac O'Raifeartaigh [93], with whom I have maintained an illuminating correspondence), we now know that Einstein made a similar statement on at least two other occasions. On the one hand, John Wheeler wrote in his 2000 book, *Exploring Black Holes: Introduction to General Relativity*, that he had personally been present when Einstein said the above words to Gamow, outside the hall of the Institute for Advanced Studies in Princeton. And another renowned physicist who will appear later, Ralph Alpher, also testified that he had heard Einstein make a similar claim. It is moreover a proven fact that Einstein never wanted to use λ again in his life. Not even when someone suggested that it might be interesting to reinstate it, since, by playing around with its value, the age of the Universe could be better adjusted to the results of observations of the oldest galaxies, which seemed at one point to greatly exceed the age of the Universe itself. As Einstein pointed out in a footnote in the appendix to the second edition of his book[4] *"The meaning of relativity"* [94]:

> If Hubble's expansion had been discovered at the time of the creation of the general theory of relativity, the cosmological term would never had been added. It now seems much less justified to introduce a term like this type in the field equations, since its introduction loses the only justification it originally had.

And it is necessary to point out an important additional fact, which almost nobody mentions: Einstein was much more radical than Friedmann and Lemaître themselves, since both always included the cosmological constant term in their models of the universe, although this was not decisive for them, as it had been in the Einstein model, since for an expanding universe solution such an additional term may play only a secondary role. If it had no reason to exist, it should not be put there under any circumstances; that is what Einstein clearly said.

In the 1930s, Hubble was involved in determining the distribution of galaxies and the spatial curvature. These data seemed to indicate that the Universe was flat and homogeneous, but there was a deviation at large

---

[4] The first book by Einstein to be produced by an American publisher, in 1921. He included new material, amplifying the theory, with each new edition.

redshifts. In December 1941, Hubble reported to the American Association for the Advancement of Science the results of a painstaking six-year investigation with the Mt. Wilson telescope. The values obtained were not consistent with the theory of the expanding universe. According to a *Los Angeles Times* article that reported on Hubble's comments, he had said [95]:

> According to the expansion model, the nebulae would not have had time to distribute themselves evenly, as the telescope shows that they are, in fact. Any explanation that attempts to circumvent what the large telescope sees does not stand. To start, the explosion would have had to take place long after Earth was created, and possibly even after the first life appeared on Earth.

The results of the telescope were correct but the big mistake Hubble made in his argument (without knowing it) was that his estimate of the distances, and therefore of the constant H_0, was wrong by a factor of more than seven! The age of the universe he estimated was just two billion years! According to Allan Sandage,

> Hubble believed that his data gave a more reasonable result with respect to spatial curvature if the redshift correction was made without assuming a recession.

From what he said even in his latest writings, he invariably maintained this same position, favoring models where there was no real expansion, and with it, the idea that redshift "*represented a principle of nature hitherto unknown.*" [96] (Fig. 5.12).

To end this section, it should be noted that, in his own work, Hubble only contributed to half of the data that led him to his famous law; namely, he did the calculation of distances. And it turns out that his values were out by almost an order of magnitude, due to incorrect distance measurements of the stars used as standard candles. In contrast, the recession rates obtained by Slipher had much smaller errors. Hubble used them with Slipher's permission (they were already public, actually, as explained above), but his name does not appear anywhere in the 1929 work. It is true that, in the 1931 paper by Hubble and Humason [81], a generous acknowledgment of Slipher's contribution does indeed appear. But it turns out that, in the two years elapsed between the two documents, Hubble's name had already been associated with the redshift-distance correlation, and the fact is that Slipher's contribution fell almost into oblivion (Fig. 5.13).

Returning to Hubble, it is fair to say that, towards the end of his life (more precisely, the year he died), he recognized Slipher's remarkable contribution,

Fig. 5.12 Albert Einstein with Edwin Hubble and Walter Adams at the Mount Wilson Observatory in 1931. Einstein visited Caltech for the first time in the winter of 1931, to discuss the cosmological implications of the theory of relativity with physicists and astronomers at Caltech and at the Mt. Wilson Observatory. Here he is looking into the eyepiece of the 100-inch Hooker telescope. Photograph Collection. Caltech Archives ID: 1.6–16. *Copyright* The Caltech Archives has not determined the copyright status of this image. Fair use

Fig. 5.13 Robert Millikan, Georges Lemaitre, and Albert Einstein at the California Institute of Technology, January 1933

in all its importance. In a letter he wrote to him on March 6, 1953 [23], he stated:

… your speeds and my distances …

Even more, Hubble recognized the great influence that Slipher's important and seminal contribution had had on his own later work, declaring that [97]:

… the first steps in a new field are the most difficult and the most significant. Once the barrier is overcome, further development is relatively simple.

In fact, Slipher was the first to realize, as I have explained before, that something very important and remarkably strange was happening with the static universe model: for how could it be static if those distant nebulae were moving away from us at such great speeds? And this was precisely what Wirtz also thought, a few years after Slipher.

In just the same year of 1929 that Hubble's work had appeared, the American mathematician and physicist Howard Robertson, based on Friedmann's two articles, published a paper in which he obtained the most general space–time metric possible under the hypothesis that the universe was both homogeneous (the same everywhere) and isotropic (the same in all spatial directions). Almost simultaneously, the English mathematician Arthur Walker obtained a very similar result. In both cases, it was found that they corresponded to one of the families of solutions that were already within those that Friedmann had obtained. But the important novelty was that Robertson and Walker managed to demonstrate rigorously that there were no more to look for: there are no other solutions or models of the universe, within Einstein's theory of general relativity, if one assumes that the universe is homogeneous and isotropic. Putting together the names of all those who contributed to the construction of the definitive metric of the space–time of a Universe satisfying the cosmological principle, this is currently called the Friedmann-Lemaître-Robertson-Walker (FLRW) metric. In reality, it is a family of metrics, since the curvature, which remains to be fixed, can still be positive, negative, or zero.

5.5 General Acceptance of Lemaître's Work of 1927

The considerable influence that Hubble already enjoyed in 1929 made the appearance of his work definitive, so Lemaître's revolutionary idea that the universe was expanding began to be taken into account by some cosmologists to explain, most naturally, the proportionality law. The first to accept this model was none other than Arthur Eddington, who had known Lemaître well since his visit to Cambridge, before his stay at MIT. It was Eddington himself who had suggested to Lemaître that he set about finding a plausible cosmological model. And Eddington had further demonstrated, by himself, that Einstein's static solution (of the equations with the cosmological constant) was indeed unstable. Moreover, by then he was already playing around with the idea of an expanding universe. So, when Hubble's work appeared in the *Proceedings of the National Academy of Sciences*, Eddington recalled that he had among his papers the article by Lemaître from 1927, in French, to which he had hitherto paid little attention. As he studied it again, more seriously, he realized this time that it was quite possibly the solution he had been searching for himself. He was pleasantly surprised to see before his eyes all the work already done by Lemaître; and the fact that the result was in complete agreement with Hubble's astronomical data and its empirical law did indeed convince him of its importance.

Eddington himself published in the prestigious journal *Monthly Notices of the Royal Astronomical Society* an extensive comment on Lemaître's 1927 article, in which he described how the latter had found a "*brilliant solution*" to the pending problems of cosmology [98]. He also helped Lemaître translate the original article, which was published in a shortened version in the same year of 1931, along with a sequel by Lemaître in response to Eddington's comments. Only part of the work appeared (*"première partie"*) [99], the one that does not contain Hubble's law, which Lemaître obtained towards the end of the paper. For a few decades, it remained a mystery why the final part of the short document, just eleven pages long, was not translated. It now seems clear that this was not due to an attempt by Hubble or by an anonymous referee in his service to hide these paragraphs. Actually, it was a personal decision by Lemaître, who did not consider those results to be important any more, after the appearance of Hubble's article, in which he had improved them by adding more data to his tables [100]. In fact, the value of the proportionality constant obtained by Hubble was 500, while Lemaître put it at 575, in the corresponding units of km per second per megaparsec. Both are a long way from the current value of between 67 and 74 (Figs. 5.14 and 5.15).

Also in 1931, Lemaître was invited to London to participate in a meeting of the British Association on the relationship between the physical universe and spirituality. There, he first proposed his next amazing idea (which I will discuss in detail in the next chapter): the universe had an origin and expanded from a starting point, an initial stage which he called the "primeval

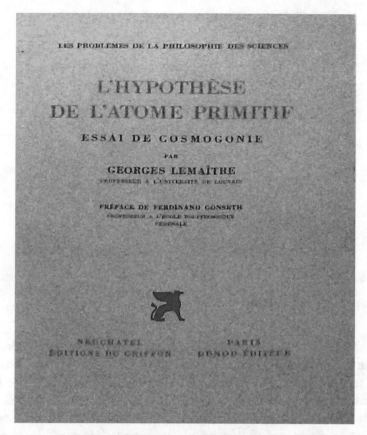

Fig. 5.14 L'Hypothèse de l'Atome primitif (The Primeval Atom—an Essay on Cosmogony); Georges Lemaître (1946); collection of the Musée L, Ottignies-Louvain-la-Neuve; Belgium. *Source* JoJan—Self-photographed, 2019. CC BY 4.0

atom." During the previous years, he had been developing this idea, which he published that same year 1931 in *Nature* and was to become his second great discovery.

5.6 2018: The Hubble-Lemaître Law

But, before continuing, I will relate below a historical event that finally recognized Lemaître's first great discovery. I will explain it in some detail, since the subject, in my opinion, is well worth it. At the XXX General Assembly of the International Astronomical Union (IAU), which was held in Vienna from August 20–31, 2018, five resolutions were proposed for approval [101]. The fifth of these, Resolution B4, addressed a proposed change of name for

Fig. 5.15 Golden cosmic egg. Work of Manaku, pahari artist, ca. 1740. *Source* https://theprint.in/pageturner/afterword/b-n-goswamy-brings-life-elusive-pahari-painter-manaku-guler/14421/

the Hubble law, recommending that, from that moment on, this law on the expansion of the universe be officially known as the "Hubble-Lemaître law." The basic arguments in favor of the resolution were that the Belgian astronomer Georges Lemaître published (in French), in 1927, the article entitled "*Un Univers homogène de masse constante et de rayon croissant rendant compte de la vitesse radiale des nébuleuses extra-galactiques*". In it, he first rediscovered one of Friedmann's dynamic solutions to Einstein's equations of general relativity which describes an expanding universe. He also deduced that the expansion of the universe explains the spectra of distant galaxies, which move towards the red by an amount proportional to their distance. Finally, he used published data on the photometric speeds and distances of galaxies to deduce the expansion rate of the universe (assuming the linear relationship he had found on theoretical grounds) (Fig. 5.16).

In the supporting bibliography, for the convenience of the reader, an extract was included from the article by David Block, "*Georges Lemaître and Stigler's Law of Eponymy*," which contains an interesting comment on the matter, due to Lemaître himself, in these terms:

"In a comment published in Nature, Mario Livio has unearthed a letter from Lemaître to W.M. Smart (dated March 9, 1931). In that document, it is clear that Lemaître himself translated his 1927 article into English and that he omitted his determination of the coefficient of expansion of the Universe (H_0) from radial velocity values available in 1927. However, in his comment,

Fig. 5.16 Presentation at the IAU General Assembly in Vienna in August 2018 of the resolution to change the name of the Hubble Law. Five Resolutions were proposed for approval at the XXXth IAU General Assembly (Vienna, August 20th–31th, 2018). They were announced and posted on the IAU web site on June 20th (see https://www.iau.org/news/announcements/detail/ann18029/). From astronomy2018.cosmoquest.org. Fair use

Livio omits a vital reference, namely a reflection written by Lemaître himself in 1950 [102]:

About my 1927 contribution, I do not want to discuss whether or not I was a professional astronomer. In any case, I was a member of the International Astronomical Union (Cambridge, 1925), and had studied astronomy for two years. One of them with Eddington and the other year at the American observatories. I visited Slipher and Hubble and listened to Hubble in Washington in 1925, when he made his memorable communication about the distance to the Andromeda nebula. While my mathematical bibliography was seriously flawed, since I did not know Friedmann's work, it was perfectly updated from the astronomical point of view. In my contribution, I calculated the value of the coefficient of expansion (575 km per second per megaparsec, 625 with a questionable statistical correction). Of course, before the discovery and study of more groups of nebulae, it made no sense to establish Hubble's Law, but only to calculate the coefficient. The title of my note leaves no doubt about my intentions: A universe with a constant mass and an increasing radius as an explanation of the radial velocity of extragalactic nebulae.

It is quite clear that, in 1950, Lemaître did not want the rich fusion he carried out of theory and observations, contained in his 1927 article, to be buried in the sands of time."

The discussion on Resolution B4 was very lively, but had to be cut to keep to the agenda of the session schedule; in particular, the subsequent closing ceremony. Therefore, some additional questions were sent by email and answered in the same way. Like the following:

Q. Should other contributors to the data used in the expansion law, such as Slipher, Leavitt, Strömgren, or others, be recognized too?

A. No, because they did not use their data or invent a new theory to discover the expansion of the universe.

In an article I published in 2019 [103], I already stated clearly that, in essence, I agreed with all the previous considerations, as they were formulated: in particular, with the last sentence, which refers to Slipher and Leavitt. No objection can be made to the wording. That is, stating that these prominent astronomers "did not use their data" (in particular, Slipher, despite being the one who calculated practically all the redshifts subsequently used by Hubble and Lemaître) or methods (this is the case of Leavitt's very important law) used extensively by Hubble as his main tool to obtain distances "to invent a new theory to describe the expansion of the universe." However, some crucial historical facts, which until now have been largely overlooked, and which some authors have been rescuing from oblivion, which contain documented opinions of the scientists involved (Hubble, in particular, on several occasions), as well as recent bibliographic studies (duly mentioned in my article), led me to formulate in it sound arguments in favor of including the prominent astronomer Vesto Slipher together with the names of Lemaître and Hubble, in order to call it the Hubble-Lemaître-Slipher (HLS) law. One should just add that Howard Robertson also obtained, as early as 1928, an expansion law, with $H_0 = 460$, using essentially the same data as Lemaître (Hubble distances and Slipher redshifts) [104].

It is true that I received some messages of support in this regard, but for the moment I will have to settle for the name that was approved in an electronic vote carried out among all members of the International Astronomical Union, among which I count myself. It yielded the following result: 78% of votes in favor of the name change, 20% against, and 2% of abstentions. In addition, quite a few of those who opposed the change did so because they opposed the reasons given. They did it rather as a matter of procedure or fear that a positive result could signify the beginning of an endless number of demands of similar kind, owing to there being so many similar historical injustices that are known in fact, in this and other fields. There is a famous principle (it already appeared above) widely known now under the name of "Stigler's eponymy law," which states [105]: "*No scientific discovery is named after its original discoverer.*" The reader may have noticed that we are encountering several clear examples of this principle along the way. I hope to give some more of them in the final conclusions of the book.

To end this chapter, I think it can be useful to summarize in a table the answer to the big question we have been trying to elucidate here[5]

5.6.1   Who was the first to discover that the Universe is expanding?

1. Vesto Slipher was the first to clearly observe, in 1914, that the universe could hardly be static. A few years later Carl Wirtz reached similar conclusions.
2. Albert Einstein masterfully established the theoretical framework, namely the theory of GR in 1915, and used his field equations, in 1917, to construct a model of the universe.
3. Willem de Sitter was the first to find, in 1917, a GR solution that describes an expanding universe, but does not contain any matter or energy. It was considered a wrong model or, at best, a toy model. Even then, it was widely used at the time, and nowadays it has become essential to model the origin and the end of our universe.
4. Alexander Friedmann was the first to say clearly, in 1922, that our universe could be expanding, since he had found a solution for GR that could be interpreted as corresponding to a universe with the characteristics of ours, and which is expanding.
5. Georges Lemaître was the first to formulate, in 1927, a scientific theory for an (eternal) universe. It directly linked astronomical redshifts (Slipher table) with the rate of expansion of his cosmological model (one of Friedmann's solutions that has a logarithmic term, which Lemaître himself had rediscovered in 1925) and saw that they are proportional to the distances (Hubble table, obtained thanks to Leavitt's law). Then, he deduced the proportionality rate two years before Hubble (who did it in 1929), and

---

[5] https://www.enciclopedia.cat/divulcat/Emili-Elizalde (in Catalan).

even explained the reason behind it. In short, he produced the first scientific theory of the expansion of our Universe, although it still lacked an origin in time.

6. In 1931, Lemaître removed the logarithmic term from his solution, reverting to the standard Friedmann solution, and concluded that the Universe had had an origin. Furthermore, he deduced that its expansion could be explained by a great explosion of an original (primeval) atom. This last point, a very important one, we will discuss below.

7. We can conclude that the Einstein-de Sitter cosmological model, of 1932, eventually marked the general recognition of the Universe's expansion.

6

The Big Bang Theory

We are now entering a crucial chapter, the most original one for the new interpretations given, and also the most difficult chapter to understand in this book. If the expansion of the universe is no longer easy to accept, getting to understand its origin in any detail is even harder. That alone can explain the enormous amount of inaccuracies and crazy absurdities that have been written, and continue to be written every day, on this issue. Many of the explanations that we can currently find in the media cling to the one that Lemaître gave 90 years ago. This is amazing. But he never said "Big Bang"! Referring again to the purpose of this book, we will delve in this chapter into the depths of the true origins of the term "*Big Bang*." When did this expression first appear? What exactly did the person who first spoke these two words mean? Why did he employ this term and not another one? And, in what context was it used?

In fact, this chapter, and with it the whole topic itself does not begin abruptly, but rather as a very smooth, absolutely logical, natural continuation of the previous chapter. Having already accepted that the universe was expanding, it was enough for Lemaître to look back in time, towards the past. It was that simple. The first point to consider, however, is that time had no origin in his 1927 model of the universe. In fact, due to the presence of a logarithmic term in his solution, the universe could reach the infinite past $(-\infty)$. However, by the year 1931—when he published his latest work in the top journal *Nature*—he had already realized that this term had no physical equivalent in the real universe. Thus, he had changed his solution of Einstein's

© The Author(s), under exclusive license to Springer Nature
Switzerland AG 2021
E. Elizalde, *The True Story of Modern Cosmology*,
https://doi.org/10.1007/978-3-030-80654-5_6

equations for precisely "the Friedman solution," the only one possible under the hypotheses of homogeneity and isotropy (as Robertson and Walker had already demonstrated by then). And in this solution, the universe does have an origin in time!

But let us now make an important point. Precisely by looking back, it was soon seen that the values of the Hubble constant, both the one Hubble himself calculated, of 500, and (even worse) the one obtained by Lemaître, of 575 km/s/Mpc, were far too large, since with these values the universe could not have been more than two billion years old. And it turned out that later studies of radioactive isotopes in rocks suggested that the Earth was at least 4.5 billion years old, thus making the universe younger than some of the objects it contained. The value of the Hubble constant had to be revised, and in fact, it has been revised a lot many times since then. We could even affirm that it is in continuous revision, being the most important of all the cosmological constants. A major correction was made in 1952, when astronomer Walter Baade discovered that Hubble had seriously underestimated galactic distances, not realizing that there are actually two different types of Cepheids. Baade's recalibration resulted in a drastic halving of the Hubble constant value. Another serious correction, made by Allan Sandage in 1958, reduced its value again, to 100 km/s/Mpc. Sandage, who had been Hubble's observation assistant, realized that what Hubble had taken to be the brightest individual stars in each galaxy were actually compact clusters of bright stars embedded in gaseous nebulae. For several decades, the value of the constant fluctuated, consistently, in the range between 50 and 100. The currently accepted value stays more or less in the middle, between 67 and 74, with a margin of error of some 5%. The associated age of the Universe, strongly constrained by many types of independent observations (we will see this later), is approximately 13.8 billion years old.

But let us get back to Lemaître. as I said before, in 1931 he was invited to London where, according to the new solution he had found (Friedmann's standard solution), he proposed for the first time that the universe had expanded from an initial moment when it had been of very small size, which he called the "primeval atom" (reminiscent of the "cosmic egg.") After publishing it in *Nature* [106], his new theory also appeared in the popular journal *Popular Science*, in December 1932 [107]. His reasoning was quite simple: he thought that, if the universe was expanding, looking back in time, it must have been getting smaller, until, as it approached the initial instant, all the matter and energy of the universe had to be concentrated in a single, extremely dense atom. Appealing to the new quantum theory of matter, Lemaître argued that the physical universe was made initially of a single particle, a "primitive atom" with a large nucleus, which disintegrated in a huge explosion, leading to the distribution of matter in the entire universe, which continues its expansion to this day. It must be said that Lemaître's proposal was initially met with skepticism from his scientific colleagues. Eddington found it very unpleasant, while Einstein thought it was completely unjustifiable from a physical viewpoint. This should come as no surprise, since Lemaître's notions of quantum physics were rather poor. What was not an obstacle for the popular acceptance of his model. Indeed, quite the opposite! His idea immediately became extremely famous, and remains so to this day.

Let us pause for a moment to reflect. Lemaître had now shown that, in addition to expanding, the universe had an origin; at which point it seemed logical to assume that a huge cosmic explosion must have occurred, which would have scattered all the highly concentrated matter and originated the expansion that we still observe today. All of this seems very reasonable and is quite easy to imagine. If we add a little 'spice' to it, a pinch of mystery in the form of the word "infinity"—on affirming that, in the beginning, the density of matter/energy was, not just very large, but *infinite*, and that the universe was reduced to a single, singular point—we then have at our disposal all the ingredients required to 'cook up' an unbeatable model: both extremely simple and highly mysterious. One only then needed to give it a spectacular and durable name, which fell from the sky a few years later: the Big Bang. With these ingredients Lemaître's model, under the name of the *Big Bang model*, was already final, unbeatable, and established forever. Well, in the popular culture, that is. That description has remained untouched for over ninety years, as if cosmology had not moved forward an inch during this time!

I have before me a truly "impossible mission": that of dismantling this simplistic construct, this caricature of the real Big Bang model. A version that appears in so many books, encyclopedias, and popular articles, in all the languages of the world, although I have to admit, with joy, that things have started to change recently. When I published my article on this subject in *Galaxies,* three years ago [108], I took the trouble to check that the description I have just given was still repeated, almost word for word, in most editions of *Wikipedia*, in almost all languages.[1] With relief I now realize that a large part of those reference pages have already been revised, eliminating crazy terms, devoid of the slightest physical meaning. Much has been gained in such a short time, in this transition from saying so much nonsense to trying

---

[1] Even today, without going any further, the Catalan edition continues to speak of "*a primitive condition in which there exist the conditions of an infinite density and temperature,*" without anyone daring to ask what that could possibly mean. Does it make the slightest sense?

to explain things, even half-heartedly, avoiding the difficult aspects. It is the least that could be asked, in a first stage, although it must be borne in mind that *Wikipedia* is the reference that is most quickly updated, contrary to other, printed texts. In what follows, we will go well beyond simply touching things up, embarking ourselves on a mission to turn this caricatural explanation of the Big Bang model upside down.

To tie up loose ends before going into the details, I should add that Lemaître and Einstein met on four occasions: the first was in Brussels, in 1927, as we have already seen; again in Belgium, in 1932, on the occasion of a series of conferences; in January 1933, in California, where both gave several seminars; and in 1935 at Princeton [109]. It was in 1933, at the California Institute of Technology, CALTECH, after an address by Lemaître in which he explained his theory, that Einstein stood up, applauded, and said that this was the most beautiful explanation of creation he had ever heard [110]. However, in the newspapers of the time, there is disagreement over what Einstein really meant. For it could be that Einstein was not referring to Lemaître's theory as a whole, but only to Lemaître's proposal that cosmic rays could be the remnants of the initial explosion of the universe (a theory that Lemaître had in fact vehemently defended). In 1933, after his approach to Einstein, Lemaître achieved his highest public recognition. Newspapers around the world called him "*the famous Belgian scientist*" and described him as the leader of the new cosmology. A photograph of Einstein and Lemaître appeared in a New York Times article with the caption: "*They have deep respect and admiration for each other.*" Of course, the fact that Lemaître was both a leading scientist and a Catholic priest explains part of the fascination he aroused in the popular press.

Lemaître was elected to the Pontifical Academy of Sciences in 1936, and served as president from March 1960 until his death six years later. In 1941, he was elected a member of the Royal Academy of Sciences and Arts of Belgium. In 1946, he published his book "*L'Hypothèse de l'Atome Primitif*" (The Hypothesis of the Primitive Atom), which was translated into Spanish the same year and into English in 1950. In 1951, Pope Pius XII stated that Lemaître's theory provided a scientific validation of Catholicism [111]. However, Lemaître stated at all times that the theory was neutral and that it did not favor or contradict religious beliefs [112]. He was always opposed to mixing science with religion, arguing that the two fields were not in conflict. In his own words:

> As far as I can see, such a theory remains completely alien to any metaphysical or religious question. It leaves the materialist the freedom to deny any transcendental Being. And, for the believer, it eliminates all attempts to become

familiar with God through it. It is consonant with Isaiah when he speaks of the God always hidden, even at the beginning of the universe.

Lemaître was the first recipient of the Eddington Medal of the Royal Astronomical Society, in 1951. In 1954, he was the first theoretical cosmologist to be nominated for the Nobel Prize in physics, for his prediction of the expanding universe. He died in 1966, shortly after learning of the discovery of cosmic microwave background radiation, which was a crucial proof of an important aspect of the hot Big Bang model, as we will see later [113].

6.1   Dark Matter

Chronologically, long before the Big Bang theory was established, the discovery of dark matter took place, so it is now necessary to make a parallel entry in the chain of logical explanations. The discovery of dark matter was a very important fact, although it went almost unnoticed for years. It raised a new question regarding our knowledge of the cosmos, which is still far from being answered even now. Already in 1884, based on the observed dispersion of the velocities of the visible stars, Lord Kelvin estimated how much dark mass the Milky Way should necessarily contain. He concluded that it was of the same order than the observable mass. Skipping other ideas on this issue, among them by Poincaré (who called it *matière obscure,* already) and others, we can actually say that it was first postulated in a clear and quantitative way by Fritz Zwicky in 1933 [114], from his detailed astronomical observations, which were extraordinarily accurate for the time, made on the galaxies of the enormous Coma Berenice cluster. Zwicky called it, in German, *"dunkle Materie"* (dark matter), because it was a type of matter that had to be there but could not be seen (Figs. 6.1 and 6.2).

This concerns a truly huge amount of matter, which current research evaluates as about six times more abundant than the ordinary matter we are able to observe (Zwicky erroneously calculated an even higher proportion). We know that it is there, without any doubt, due to its gravitational effects. These include, in particular, anomalies in the rotation curves of galaxies, their effects as gravitational lenses (according to Einstein's theory of general relativity), and others. We do not yet know what this matter is made of, for it has escaped any type of direct detection, in spite of the many attempts undertaken in different laboratories worldwide (Fig. 6.3).

Fritz Zwicky was one of those great astronomers who sadly remained unrecognized by the general public. Today his name is sometimes forgotten when talking about dark matter. In addition to discovering the existence of

Fig. 6.1 Fritz Zwicky (1898–1974). The picture appears on the website of the Fritz Zwicky Stiftung (the Swiss Fritz Zwicky Foundation at: http://www.zwicky-stiftu ng.ch/). Fair use

Fig. 6.2 Elements of the Coma Cluster of galaxies. Three of the galaxies seen in this image are IC 4040 (left edge), NGC 4889 (center-left), and NGC 4874 (center-right). The bright star in the upper right is HD 112887. See image annotations. Credit Line and Copyright Adam Block/Mount Lemmon SkyCenter/University of Arizona, 2012. CC BY-SA 3.0

Fig. 6.3 Memorial plaque on the house in Varna where Fritz Zwicky was born. His contributions to the understanding of neutron stars and dark matter are mentioned explicitly. *Source* PetaRZ—Own work, 2008. CC BY-SA 4.0

dark matter, he made many other far-reaching discoveries. For example, he understood the importance of the gravitational lensing effect of the big mass concentrations in the cosmos. These deflect the paths of the light rays that pass near them, in agreement with Einstein's equations of general relativity. In 1937 Zwicky calculated the value of this effect (which he called the Einstein effect) to very good accuracy and predicted the enormous importance it was going to have in astronomy [115]. This phenomenon was not confirmed by precise direct observations until 1979. Dark matter is nothing more than its name suggests: matter that cannot be seen. Its gravitational behavior is exactly the same as that of ordinary matter. In this sense, it is not a strange thing at all. In fact, it was believed for some time that it could be made of large massive objects, the so-called massive astrophysical compact halo objects (MACHOs), such as Jupiter-sized planets, that are very difficult to observe in distant star systems. Massive and compact astrophysical objects, since they are not luminous, are extremely difficult to detect. Aside from planets like Jupiter, this option also includes certain neutron stars, relatively small black holes, and brown dwarfs. Very recently, primordial black holes have been proposed as candidates. Still, it turns out very difficult to explain, in its entirety, the enormous proportion of dark matter in the universe and its distribution, which we now know more and more precisely. Another possible component that has been analyzed is neutrinos, which have been found to have a mass, albeit a very small one. Despite the huge amount of neutrinos, their contribution does not seem to solve the problem either (Fig. 6.4).

Current astronomers who are very famous for their studies of dark matter, since they rescued this important problem from an oblivion that had lasted for more than forty years, are Vera Rubin (who died in 2016), Norbert Thonnard, and Kent Ford, for their work published in the 1970s [116]. It is worth mentioning, for the most seasoned readers, that alternative explanatory proposals have emerged that eliminate the existence of dark matter altogether. They are based on a large-scale modification of Newton's laws. Vera Rubin did sometimes vehemently defend some of these alternative proposals. But the vast majority of researchers in dark matter are looking for the so-called *wimps* (weakly interacting massive particles). They do it in large projects such as [117]: the XENON Dark Matter Project (XENON1T), an experiment

Fig. 6.4 Vera Rubin (1928–2016). Photograph NASA—These pictures of posed groups from the NASA Sponsors Women in Astronomy and Space Science 2009 Conference

Fig. 6.5 From left, W. Kent Ford, Jr., Norbert Thonnard, and Vera Rubin in 1981. Courtesy Carnegie Institution for Science, permission granted

located on the Gran Sasso mountain in central Italy; the European Underground Rare Event Calorimeter Array (EURECA), to be built in the Modane Underground Laboratory, in the Fréjus tunnel between France and Italy, and which is the deepest underground laboratory in Europe; the Axion Dark Matter Experiment (ADMX), an American project of the University of Washington, in Seattle (USA); and the LZ Dark Matter Experiment, consisting of 10 tonnes of liquified xenon to detect faint interactions between galactic dark matter and regular matter. High hopes are placed, in particular, on the XENON1T project, sponsored by CERN, the world's largest particle physics center based in Geneva, Switzerland, which not long ago discovered the Higgs boson. This in turn had fundamental implications in cosmology, but that would take us beyond the scope of this book. We should be very proud of all these projects and achievements; even more, of the prospects they open for the future (Figs. 6.5, 6.6 and 6.7).

6.2   Fred Hoyle

As pointed out, Lemaître's model of the primitive atom was met with some reluctance from the beginning. It should be mentioned that the early specialists in quantum physics soon realized that it was not consistent. It immediately met resistance from nuclear physicists, such as Walter Adams, Theodore Dunham, George Gamow, Ralph Alpher, Robert Herman, and, above all, with Fred Hoyle, who rather quickly understood that this model wobbled. Let us get this right: not the fact that the universe had an origin,

Fig. 6.6 Norbert Thonnard, Vera Rubin, and W. Kent Ford Jr. gathered once again in May 2013 during a ceremony celebrating the historical importance of their work on the rotation curves of galaxies (Janice Dunlap/DTM). Courtesy Carnegie Institution for Science, permission granted

and that there must have been some impulse or instantaneous force to set these huge masses in motion, according to their current state of expansion. This idea was reasonable, but it was the precise explanation in the form of a primitive atom, concentrating all the masses and energies of the universe.

Fred Hoyle (1915–2001) was an Englishman, a nuclear astrophysicist who is now known primarily for his theory of stellar nucleosynthesis; in other words, for being the discoverer that *we are all stardust*. Wow! Imagine! Has anyone ever come up with a more amazing concept than this? Like several other nuclear physicists, Hoyle had realized that, under the conditions of the models for the early universe (which had already gone much beyond the cosmic egg idea[2]), the production of heavy atoms would never have been possible. At best hydrogen, helium, and perhaps a small amount of lithium could originate there. Hoyle was the author of the first two published research papers on the synthesis of chemical elements heavier than helium by nuclear reactions in stars. In the first of these [118], from 1946, he showed that under certain conditions the cores of the stars could evolve to temperatures of billions of degrees, which was much higher than the temperatures considered for the cores of the stars of the main sequence. He showed that, at

---

[2] Bridging the gap, we could say that that model was reminiscent of some proposals that emerged many centuries ago, such as the allegory of the cosmic egg, an explanation of the origins of the Universe and the Earth that is found in various cultures, such as Egyptian, Hindu and Chinese, and which later reappeared in ancient legends of northern Europe (Fig. 5.15).

Fig. 6.7 The ADMX magnet being installed at the University of Washington. Although installed below the floor, the detector is in a surface laboratory. *Source* Lamestlamer—Own work. Created: 25 May 2011. CC BY-SA 3.0

such high temperatures, iron could become much more abundant than other heavy elements, due to the thermal balance between nuclear particles, which explained the large natural abundance of this element on our planet. This idea would later be called the *e process* [119] (Fig. 6.8).

In his second groundbreaking publication, Hoyle showed [120] that the elements between carbon and iron could not be synthesized by such equilibrium processes. And he attributed the production of those elements to specific nuclear fusion reactions between constituents that abound in concentric layers as massive stars evolved into pre-supernovae. This image, surprisingly modern for the time, is the accepted paradigm today for nucleosynthesis of these chemical elements, which takes place in supernova explosions. In the mid-1950s, Hoyle led a highly experienced group of theoretical and experimental physicists at Cambridge, including William Fowler and Margaret and Geoffrey Burbidge. This group systematized the basic ideas about how all the chemical elements in our universe (except for the lightest ones, hydrogen and helium) were created, a field known as stellar nucleosynthesis. In 1957, the group wrote a very famous article, known as B^2FH (by the initials of the four authors [108]), that laid the foundations for this discipline, which continues to be of enormous importance today. In particular, the relative weight of the neutron capture *r* and *s processes* in nucleosynthesis was established in this paper.

Fig. 6.8 Photo of Fred Hoyle (1915–2001). Fair use

Concerning Lemaître's model, the first to improve it and give it a much more rigorous physical sense was George Gamow, a physicist that has already appeared, of whom I will say much more later, and who actually became quite famous. He had been a student of Friedmann and, towards the end of the 1930s, his interest moved towards cosmology, so he decided to improve the original model of Friedmann and Lemaître. Indeed, Gamow reformulated from top to bottom the very rough idea of the primitive atom or cosmic egg and managed to endow this model with a much more precise physical meaning, already in accordance with the knowledge of nuclear physics of the time. This was actually the model that Fred Hoyle criticized in his famous BBC "Big Bang" broadcast. But let us pause a bit.

During World War II, Hoyle worked in Britain on the radar project with Hermann Bondi and Thomas Gold. Although Hoyle never received the Nobel Prize, his colleague William Fowler, who won it in 1983 for his work on stellar nucleosynthesis, acknowledged that:

The concept of nucleosynthesis in stars was first established by Hoyle in 1946.

Hoyle had always considered the idea that the Universe had a beginning as pure pseudoscience, mere arguments to introduce a creator into it,

… because it is an irrational process and cannot be described in scientific terms;
… a mere belief in the first page of Genesis.

In 1948, Hoyle, Bondi, and Gold published their steady state theory, which was to become very famous and involves a "*creation*" or "*C field*." To be precise, it was Hoyle who first published an article on this theory and also the one who later gave it a mathematical form. Bondi and Gold's intention, in their work of the same year, was something philosophically more ambitious: to extend the cosmological principle as far as possible, to encompass what they called "*the perfect cosmological principle*," according to which the Universe always has the same properties, anywhere and at any time.

But, before continuing along this path I am forced, once again, to go back, in order to refer to a handwritten note by Einstein, drawn up in 1931 (most probably in January) and only recently discovered. In it, he already considered a model of a similar kind, 17 years before HBG. The note has four pages and was never published. Its title is "*Zum kosmologischen Problem*" (*On the cosmological problem*) and is preserved in the archives of the Hebrew University of Jerusalem [121]. Einstein's reasoning goes as follows:

If you consider a physically limited volume, the particles of matter will continually leave it. In order for the density to remain constant, it is necessary that new particles of matter are continuously produced in the volume, from the space itself.

Almost certainly, Einstein's note was never known to Hoyle, Bondi, and Gold, but it describes, in essence, the same physical principle as their theory of the steady state. This is also the same principle on which all current models of cosmic inflation are based. It is impossible to describe it better than in the above words by Einstein. But he abandoned his manuscript because he found an error in it and was eventually unable to provide a reasonable mechanism to perform the conversion of space into matter, as he needed. He intended to carry out this transmutation using the cosmological constant, but finally realized, in a last revision that he made of his note before sending it for publication, that this was not possible in practice. The note remained in a drawer and Einstein ignored it thereafter. With hindsight, we know the reason for his failure: he was unable to see that he needed an additional term with a field that could create matter from the geometry, as in fact Hoyle, Bondi, and Gold did in their model. However, when examined in detail, their mechanism was not very precise either, and more importantly, no evidence was ever found then or now that such a creation of matter in galaxies could actually take place. Tying up loose ends, it is most likely that this was Einstein's last attempt, in early 1931, to rescue (albeit in a rather different fashion) his stationary universe model by using the cosmological constant. I am now personally convinced that it was precisely this failure that finally decided him to expel λ so abruptly and radically from his equations, that very same year and for the rest of his life.

It should be remembered that, even in the 1940s, there were many physicists and astronomers who continued to hold the conviction that the stationary model of the Universe was the right one. Interestingly enough, there was never any doubt about Hubble's empirical law, which was always considered correct by everybody. Therefore—as we have just seen, masterfully explained by Einstein—to compensate for the loss of density of matter/energy due to the distant galaxies moving away from us, a term had to be introduced in the equations to create matter/energy out of nowhere. Well, not out of nowhere, but from space itself (as Einstein put it) or, if you like, from the geometry or fluctuations in the fabric of space. In short, from the mathematical part of the GR equations. Einstein had tried it through the cosmological constant, without succeeding. Hoyle, Bondi, and Gold (half-)achieved this through a "creation" or "C-field," which generated matter inside galaxies in distant regions of the cosmos, at a relatively smooth rate, just enough to compensate for the volume increase of the visible universe. For it must be taken into account that in this model space did not expand at all: these authors completely denied Lemaître's model and Friedmann's solution. They held onto the idea that the expansion of space as a whole was a simple mathematical trick that had nothing to do with reality. The question that now needs to be clarified is: How is it possible to generate matter or energy out of a reference system, from the mathematics of space–time? That is where Einstein's magnificent GR equations come into play. I repeat again that what I am now going to explain is exactly the same physical principle that allows for the creation of the quark-gluon plasma in current cosmic inflation theories.

Unfortunately, many people today are unaware that the concepts of a universe of "*zero total energy*" or of a "*free lunch*," that is, a universe with zero total energy, maintaining the energy balance both at its origin and in its later evolution, are not concepts invented by Alan Guth, Andrei Linde, or Alexander Vilenkin, nor by any of the other famous inflationary physicists. These concepts were already explained in an absolutely masterful way, for example, in the famous 1934 book by Richard Tolman, "*Relativity, Thermodynamics and Cosmology*" [122]. There, one finds categoric sentences like this:

> … a closed universe can have total energy equal to zero. All of its mass/energy is positive but its gravitational energy is negative, and they can cancel each other out, leading to a universe of zero total energy.

In Alan Guth's brilliant physics classes at MIT—which I have had the pleasure of attending—this statement is now called the "*Miracle of Physics No. 2*", which reads as follows: "*Energies are not always positive; the gravitational field has negative energy.*" It is preceded by the "*Miracle of Physics No. 1*", which Guth explains in the same lesson, and which is in fact what interests us now: "*Gravity can be repulsive.*" According to general relativity, it is possible to maintain the energy balance (or principle of energy conservation) all the way, as follows. A positive amount of matter/energy can be generated as long as an equivalent amount of negative space–time pressure (also known as inflation) is produced, for example, corresponding to an expansion of the reference system (in mathematical terms), or what comes to the same, of the fabric of space (as physicists usually put it). Problem solved! Inflation manages to find a specific mechanism for doing this: converting an expansion of space–time into matter. This possibility is already implicit in Einstein's field equations, but he himself was not able to carry it out in a concrete way, as we saw above, and Hoyle, Bondi, and Gold only succeeded in part: their mechanism was incomplete and without any physical evidence.

To better understand this most important point, let us turn to the second Friedmann equation:

$$\frac{\ddot{a}}{a} = -\frac{4\pi G}{3}\left(\rho + \frac{3p}{c^2}\right) + \frac{\Lambda c^2}{3}.$$

Without pausing to make a detailed, more rigorous analysis, it is enough to observe as follows, in order to understand the concept. On the left, we have the cosmic acceleration: a is the so-called scale factor, a typical length large enough for the expansion of the universe to be noticeable (typically 100 Mpc, or so); $\ddot{a}$ is its second derivative with respect to time. The minus sign, together with the density of matter ρ, indicate that the masses of the universe slow down its expansion (Newton's law). Einstein's cosmological constant, Λ, can contribute to the acceleration of the cosmos (if it is positive) or to its deceleration (if it is negative). And we shall now consider the pressure term p for a moment. The difference between GR and classical mechanics lies in this simple-looking term. This pressure does *not* refer to the pressure exerted by a gas, or by a tremendous explosion that occurred somewhere in the cosmos. Nothing like that could have the least bit of meaning here, where we are dealing with the spacetime geometry. It is the pressure corresponding to the compression (positive pressure) or expansion (negative pressure) of space itself, of the very fabric of the cosmos. The really important contribution is what lies inside the round brackets in the equation: matter and pressure *go together*, they are interchangeable one for the other. We can increase the mass of the universe (that is, create matter, which is always positive) at the cost of adding the equivalent amount of negative pressure to the fabric of space, and the equation will remain unchanged. The negative gravity, of which Guth speaks, can be seen as an expansion of space (negative p), when compared with the ordinary gravity of the matter term ρ. And, when there is as much matter as expansion, the situation can arise where the total energy of the universe is exactly zero. It is difficult to go beyond what I have just explained, in these few lines. Please read this paragraph once more, if necessary.

I am afraid once again that we have come too far along this path, and must now return to the point where we left our narrative. By the end of the 1940s, Hoyle, who had long since demonstrated his capacities as a nuclear physicist, already perfectly understood that Lemaître's primitive atom or cosmic egg did not bear scrutiny. And neither did the substantial improvement of the model worked out by George Gamow. It was absurd to imagine that all the mass and energy of the universe could be concentrated at the tip of a needle. On the other hand, he knew perfectly well that the generation of the atoms of almost all chemical elements, apart from the two or three lightest ones, required the much higher energies and much more elaborate processes involved in stellar evolution. When he compared Gamow's model with his own steady state model, in which matter was generated from multiple but small expansions of space, he understood that Gamow had no option but to generate *all the matter at once*, at the very beginning of the universe, in a tremendous expansion of the fabric of space. To create all that huge amount of matter and instantly! This was quite inconceivable.

6.3   The Big Bang

That is exactly what Hoyle had in mind, and it is word for word what he said in a now famous broadcast on the BBC Third Programme for radio on March 28, 1949:

> [Gamow's model implies that] … all matter in the universe was created in one Big Bang at a particular time …

And he said these words (according to the chroniclers) with a derogatory intonation, to make clear that this fact seemed to him absurd, indeed completely impossible. This (and none other!) is the precise meaning of the words "*Big Bang*," coming out of the mouth of the person who invented this concept, now so famous. One more thing must be said: going back to the origins of the origin of the term, it turns out that, contradicting what is read in so many places, Fred Hoyle was actually not the first to utter these two words (Fig. 6.9).

Since the discovery that the most distant spiral galaxies were receding at high speeds, many astronomers were convinced that there must have been an initial impulse responsible for bringing about the expansion of all celestial masses. John Barrow personally communicated to me that, in particular, among the Cambridge astronomers and cosmologists, where Arthur Eddington was particularly prominent, the terms *Bang* and *Big Bang* had often been used in the 1930s to designate this original force. Thus, these words had the sense of an initial impulse, or huge thrust, produced by some kind of cosmic cataclysm or other similar instantaneous explosion, necessary to explain the scattering of the galaxies, already observed for the first time by Vesto Slipher in 1914. However, Eddington, for one, did not believe that this could have actually happened. In his book "*The Nature of the Physical World*," he wrote [123]: "*As a scientist, I just do not think the Universe started with a bang.*" In short, Hoyle did not therefore invent the term "Big Bang," but

Fig. 6.9 From left to right, Thomas Gold, Hermann Bondi, and Fred Hoyle at a conference held during the 1960s. Fair use

he did give it a completely new, radically different meaning, with a scientific basis rooted in the deepest principles of Einstein's general theory of relativity. A meaning that very few, outside the true specialists, are capable of understanding properly *even today*, after more than seventy years. Only this can explain why a concept so well formulated and expressed so precisely and with the utmost rigor from the outset, is described today in popular references in such wrong and crazy terms.

It is thus clear that Hoyle gave the term "Big Bang" a completely different meaning than the one it had had in Cambridge until then. From being an ordinary impulse, a thrust that simply put the previously existing masses in motion, it became a creation impulse, a huge expansion of the very fabric of space, an enormous negative pressure that would allow the creation of the formidable mass and positive energy of the entire universe, out of nothing, in a single creative blow. That, and many more things as we have seen, is what Einstein's GR theory *makes possible*, in principle. But, what *precise* mechanism could be responsible for such a huge expansion of space? None at all, according to Hoyle.

However, sometimes impossible things do happen. Exactly thirty years later, Alan Guth, an American Ph.D. in theoretical physics who was about to finish his last post-doc contract, which would have left him on the street (as is so often the case today in my own country, unfortunately), faced with the urgent need to do something absolutely extraordinary, was able to give birth to a theory which he called cosmic inflation, which not only did what Hoyle considered absolutely impossible, but also solved in one go all the other cosmological problems (causality, the horizon issue, the absence of magnetic monopoles, etc.) that had accumulated over several decades, and affected all models of a Big Bang universe considered until then. The physics on which he based the theory of inflation was exactly the same as Hoyle and his colleagues, and Einstein himself, had used in their steady state models: neither more nor less (the reader will have guessed it already) than general relativity.

Until his death, Hoyle fiercely defended the similarity of his theory with that of cosmic inflation, going so far as to affirm that the latter was little more than what he and his collaborators had already developed many years before. This is not true, far from it. The grounds, the physical principles, are exactly the same; but inflationary models, which now number in the dozens, are far more elaborate and predictive.

I will not speak here of the dark side of Fred Hoyle. I will only mention that his enormous influence at Cambridge had for years prevented students from learning about the theory of the expansion of the universe, even long after cosmic background radiation had already been detected and fully confirmed it. Hoyle promoted the hypothesis that the first life on Earth started in space, spreading through the universe through panspermia, and that evolution on Earth was influenced by a constant influx of viruses coming through comets. He even claimed that outbreaks of many diseases were of extraterrestrial origin, including the 1918 influenza pandemic, certain polio outbreaks, and the mad cow disease. All this probably contributed to his never receiving the Nobel Prize, which his work had amply deserved. In an editorial article in *Nature*, John Maddox wrote on the occasion of the 1983 Nobel Prize to Fowler that it was a real shame that Hoyle had not received it, too. In turn, Hoyle got many other distinctions. He was an enormously prolific author and featured on many radio and television programs. He died in 2001 at the age of 86 years.

Anyway, the fact is, that Hoyle is now often remembered as the scientist who proposed the discredited steady-state theory of the universe, which was ultimately proven to be a wrong model. John Gribbin, in Hoyle's obituary, to which he put the beautiful title of "*Sturdust Memories*," [124] wrote:

Everyone knows that the rival theory, the Big Bang, won the battle of cosmologies, but few (not even astronomers) appreciate that the mathematical formalism of the now-favored version of Big Bang, called inflation, is identical to Hoyle's version of the Steady State model.

In truth, Hoyle was the one who invented the stardust concept, the man who proved that we are all made of stardust, since our bodies contain a lot of elements that are no more than ashes from stellar explosions. Pure poetry, like the beautiful title of Hoyle's obituary. However, I have to agree with Alexei Starobinsky[3] that to assert as Gribbin does that the mathematical formalism of inflation is the same as that of Hoyle's steady state is to take the analogy too far. What is, in fact, identical is the underlying physical law, which allows the creation of matter from the negative pressure of the fabric of space, preserving the principle of energy conservation at all times, until the end of the process, as so clearly explained in Tolman's 1934 book [111]. Ultimately, everything is just pure general relativity, interpreted as it should be.

One final but important observation is that Hoyle invoked the Big Bang (his 'inflation') for one and only one purpose, namely, to create matter and energy, and not for any other of the crucial reasons that eventually led to its formulation by Alan Guth, in his famous 1981 article [125], viz., the problems of the horizon, flatness, causality, and magnetic monopoles. This point must be emphasized, although it should have been clear already from our previous discussion.

So far, I have explained the precise meaning that the term "Big Bang" had for Hoyle, when he pronounced these two words. Later they would be applied, as we will see, in very diverse contexts. Perhaps the most common context today is that of the Big Bang singularity. But, incredible though it may seem, historically, the first to take possession of the name, despite the fact that Hoyle used it in a rather pejorative tone—as an absurd, unimaginable possibility—was George Gamow in his improved version of Lemaître's model. My opinion is that this was because nobody understood at the time what Hoyle meant, despite the fact that his words were very clear, as I have explained. As already put forward, and we will discuss later, the baton he passed on was not picked up until 30 years later by Alan Guth, who took up Hoyle's 'impossible' challenge in a brilliant way with his theory of inflation. All theoretical cosmologists will agree on this point (Fig. 6.10).

6.3.1   George Gamow

The first to definitely improve Lemaître's model and give it a physical meaning, under the name of "*the Big Bang model*," was George Gamow, already famous for various reasons. Given the imaginative character that he always showed, it was hardly surprising that he would choose precisely the derogatory name that Fred Hoyle had given this model to denigrate it. Born in 1904, in Odessa (Ukraine, then part of the Russian Empire), Gamow studied at the University of Leningrad where he was a student of

---

[3] Starobinsky made this comment to me at the end of one of my talks in Kazan, Russia.

Fig. 6.10 George Gamow (1904–1968). *Source* https://www.colorado.edu/physics/eve nts/outreach/george-gamow-memorial-lecture-series. Fair use

Friedmann and a friend of Lev Landau. After a brilliant career in Russia, where in 1933, at the age of 28, he was already elected a corresponding member of the Academy of Sciences, he fled his country with his wife, taking advantage of his participation in the 7th Solvay Conference, held in Brussels. He was supported by Marie Curie and other physicists, who helped him extend his stay at European universities. The following year, he went to George Washington University, where he was appointed professor. Once there, he immediately recruited Edward Teller, who came from London to collaborate with him. They worked together on the beta decay theory, as did Ralph Alpher, Robert Herman, and Hans Bethe. Together, they published the famous paper now known as Alpher-Bethe-Gamow [126], a fact that is usually exhibited as clear proof of Gamow's imaginative character. Towards the end of the 1930s, his interest turned to cosmology, in particular to what is called the Big Bang nucleosynthesis problem. He was the one who rescued the Friedmann and Lemaître models from oblivion, and completely reformulated the cartoonish (I would even call it Dalinian) idea of the primeval atom or cosmic egg, to endow it with a precise physical sense, in accordance with the knowledge of nuclear physics of the time. As I have already said, he took the initiative of renaming it as the Big Bang model, a name that has remained ever since. He published about twenty works on cosmology, before his interests turned again, this time to the study of the genetic code, as is magnificently described in an article in *Nature* by Gino Segrè entitled "*The big Bang and the genetic code*" (Figs. 6.11 and 6.12).

In 1948, Gamow published an article that described the production of protons and deuterons from thermal neutrons, in which he also obtained the density of matter at the beginning of nucleosynthesis, and from this, the mass and diameter of the first galaxies [127]. In 1953 he obtained similar

Fig. 6.11 A group of scientists who participated in the 7th Solvay Conference, Bruxelles (October 1933). Seated (left to right): Erwin Schrödinger, Irène Joliot-Curie, Niels Henrik David Bohr, Abram Ioffe, Marie Curie, Paul Langevin, Owen Willans Richardson, Lord Ernest Rutherford, Théophile de Donder, Maurice de Broglie, Louis de Broglie, Lise Meitner, James Chadwick. Standing (left to right): Émile Henriot, Francis Perrin, Frédéric Joliot-Curie, Werner Heisenberg, Hendrik Anthony Kramers, Ernst Stahel, Enrico Fermi, Ernest Walton, Paul Dirac, Peter Debye, Nevill Francis Mott, Blas Cabrera y Felipe, George Gamow, Walther Bothe, Patrick Blackett, M.S. Rosenblum, Jacques Errera, Ed. Bauer, Wolfgang Pauli, Jules-Émile Verschaffelt, Max Cosyns, E. Herzen, John Douglas Cockcroft, Charles Drummond Ellis, Rudolf Peierls, Auguste Piccard, Ernest O. Lawrence, Léon Rosenfeld. *Source* Benjamin Couprie—https://research.archives.gov/id/7665680?q=rutherford%20ernest

Fig. 6.12 Ralph Alpher (1921–2007) in 1950. *Credit* Alpher Papers—https://link.spr inger.com/article/10.1007/s00016-012-0088-7. CC BY-SA 4.0

results based on another determination of the densities of matter and radiation at the time when the two values became exactly equal [128]. In this very work, Gamow determined the density of the remaining background radiation, from which he predicted, before anyone else, that it should at present be detected as a background temperature of 7 K; a value that is only slightly more than twice the current one obtained for the cosmic microwave background (CMB), discovered in 1965. That was a very important prediction, although it was not formulated in his paper with such great clarity and went completely unnoticed. To Gamow's great credit, I can say that describing everything he did, even in summary, would take me more than one page. The anecdotes to which he gave rise are numerous. I will limit myself to adding only that he wrote several popular books that gave him extraordinary fame as a high-level scientific popularizer. In particular *"One, two, three, ... infinite"*, which is one of the biggest bestsellers in the history of popular science literature. He died at the age of 64, in 1968 (Figs. 6.13 and 6.14).

Fig. 6.13 Grave of George Gamow in Green Mountain Cemetery, Boulder, Colorado, USA. Personal photograph. No rights claimed

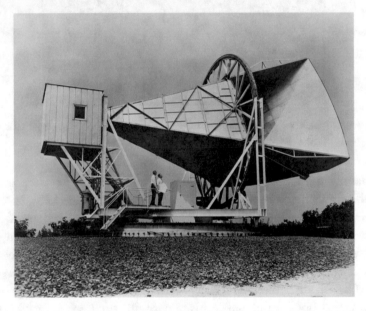

Fig. 6.14 The Holmdel horn antenna with which Arno Penzias and Robert Wilson discovered the cosmic microwave background. The antenna was constructed in 1959 to support Project Echo—the National Aeronautics and Space Administration's passive communications satellites, which used large earth-orbiting aluminized plastic balloons as reflectors to bounce radio signals from one point on the Earth to another. NASA, restored by Bammesk—Original version at Flickr: NASA on The Commons. This file was derived from: Horn Antenna-in Holmdel, New Jersey.jpeg. Created: 1 June 1962

6.4 The Cosmic Microwave Background Radiation

In 1963, Arno Penzias and Robert Wilson were working at the Bell Labs in New Jersey, recalibrating a reflective antenna which had already been used for several years and which they wanted to transform for use in radio astronomy. Despite the fact that much more powerful radio telescopes existed in some places at that time, that modest seven-meter horn-shaped reflector had unique characteristics for the high-precision measurements they wanted to carry out at the 21 cm wavelength, at which the galactic halo would be bright enough to detect it and at which the line corresponding to neutral hydrogen atoms would be observed. They particularly wanted to observe the presence of hydrogen in galaxy clusters (there is a very precise description in Robert Wilson's Nobel Prize lesson [129]). After a series of measurements carried out over several months, they were unable to eliminate a very weak but persistent noise which, translated to temperature, was equivalent to about 3 K, and which was exactly the same in all directions, at any time of day or night. The first thing they considered was the possibility that some terrestrial source was responsible for the noise, so they focused the antenna in various directions, in particular pointing it towards New York; but the variation was always insignificant (Figs. 6.15 and 6.16).

They also took into account the possible radiation from the galaxy, as well as all types of radio emissions, but nothing explained the background

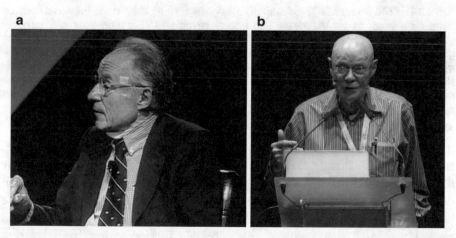

Fig. 6.15 **a** Arno Penzias (1933–). *Source* Kartik J at English Wikipedia. Created: 17 July 2007. CC BY-SA 3.0. **b** Robert Wilson (1936–) at Starmus 2016. Tenerife, Canary Islands. *Source* Victor R. Ruiz de Arinaga, Canary Islands, Spain. Created: 27 June 2016. CC BY 2.0

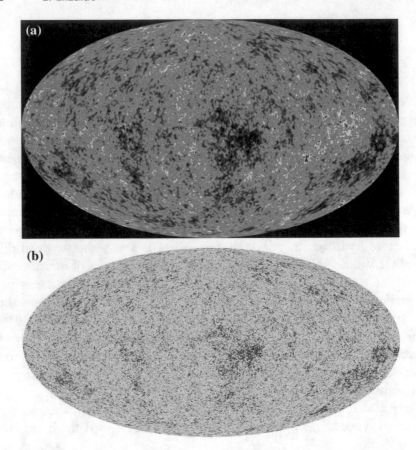

Fig. 6.16 CMB images. **a** The detailed, all-sky picture of the infant universe created from nine years of WMAP data. The image reveals 13.77 billion year old temperature fluctuations (shown as color differences) that correspond to the seeds that grew to become galaxies. The signal from our galaxy was subtracted using the multi-frequency data. This image shows a temperature range of ±200 μK. *Credit* NASA/WMAP Science Team WMAP # 121238. Image caption 9 year WMAP image of background cosmic radiation (2012). **b** The Cosmic Microwave Background as seen from the Planck satellite. *Credit* ESA https://www.esa.int/ESA_Multimedia/Images/2013/03/Planck_CMB Created: 5 August 2021, CC BY-SA 4.0

noise. They were already desperate and the anecdote is famous that, one lucky day, they realized that the antenna was partially covered by a layer of pigeon debris; they were very glad to have finally found the solution. However, the joy was short-lived: after cleaning the antenna, the signal was still there! And even when, later, they recoated the antenna's surface with a fresh layer of aluminum.

So, a whole year passed. Simultaneously, around the same time and only 60 km away, in Princeton, Robert Dicke, James Peebles, and David

Wilkinson were working to formulate a theory in which they considered the expected characteristics of microwave radiation reaching us from a universe—very dense in origin and possibly pulsating—under conditions similar in fact to those of the Big Bang model. It was Bernard Burke, a professor at MIT, who spoke to Penzias about the work of Peebles and his teammates. Collaborating together, they tied up the loose ends and, finally, during 1964 they decided to write two articles: those of Princeton with the theoretical model and Penzias and Wilson with the observations they had made with the antenna. They were published in 1965, in the same issue of the Astrophysical Journal. The theorists had already realized that there was in fact a good chance that Penzias and Wilson had detected the blast wave from the Big Bang itself!

Be that as it may, the final confirmation—which dispelled all possible doubt—of that great cosmological discovery, of colossal importance, was still delayed for a few years. And it should be said that the first additional evidence was in fact obtained by rescuing from oblivion some indirect measures that Walter Adams and Theodore Dunham had carried out thirty years earlier [130]. Reanalyzed in 1966, these data led to the firm the conclusion that they had already detected (although without giving it the importance it deserved) a background radiation of about 2.5 K. This value was further improved to 2.8 K in an article published in *Nature* in 1966, already very close to the value of 2.725 K that is now known with great accuracy.

And regarding the underlying theory, something very similar happened: reviewing the literature, it was discovered that the first theoretical model describing the Big Bang radiation was not the one we have just mentioned, but another published sixteen years earlier, proposed for the first time by George Gamow in 1948 and improved by Ralph Alpher and Robert Herman in 1949 [131]. Those authors are now recognized as the first to predict the Big Bang microwave background radiation, for which they calculated a value of between 5 and 7 K. A value that they then blotted out, in a later calculation, from 1967, bringing it to 28 K; it is believed that they were wrongly intending to make the effect more easily measurable. This discovery earned Penzias and Wilson the 1978 Nobel Prize in Physics. At the same time, radiation from the Big Bang conclusively and definitively ruled out Hoyle, Bondi, and Gold's steady state theory. The work of Peebles and collaborators was rewarded by the Nobel Prize in Physics in 2019 (Figs. 6.17 and 6.18).

The original Big Bang theory had to be modified in the early 1980s in order to resolve some very important discrepancies that it had in relation to the most accurate observations of the Universe, especially regarding the description of the first second that followed its origin. An inflation stage was

Fig. 6.17 Robert Dicke (1916–1997). *Source* WP:NFCC#4. Fair use

Fig. 6.18 David Wilkinson (1935–2002). *Source* NASA—http://map.gsfc.nasa.gov/mission/wilkinson.html

incorporated, in which the expansion was incredibly huge: in a tiny fraction of a second, the Universe went from having the volume of a pea to that of the current Milky Way. In addition, it was during this moment that, starting from a practically empty (quasi-de Sitter) universe, the generation of the quark-gluon plasma occurred, once inflation ceased. Today, inflation embodies what Hoyle had considered completely impossible: his Big Bang responsible for the creation of the matter and energy of the universe,[4] giving it a physical nature and a precise formulation.

But before talking about the theory of cosmic inflation, I think it my duty to mention a previous study, to prepare the way. Spontaneous creation of something even as small as a molecule is highly unlikely. However, in 1973, an assistant professor at Columbia University, Edward Tryon, suggested that the entire universe could have arisen this way. In his article "*Is the Universe a fluctuation of the vacuum?*" he stated: "*I launch the modest proposal that our Universe is simply one of those things that happen from time to time.*" That year I had finished my degree in physics and I still remember the impression that this article had on me. I saw it first as a preprint, one of the first I ever had in my hands. My curiosity was high. However, I remember that many scoffed at the idea. If a molecule that existed for just a brief moment, emerging virtually out of the vacuum, was tremendously unlikely—physicists reasoned—a whole universe coming out of nowhere … that was impossible.

6.5   Alan Guth

There is a fascinating description, due to Alan Guth himself and published in the Boston Globe—which I once took from his website—about how, finding himself in a very desperate situation, he arrived, after a sleepless night, at the concept of cosmic inflation.

Alan Guth grew up in Highland Park, New Jersey, occasionally helping his father in the family dry cleaning business. He was a very intelligent boy and his teachers did not know what to do with him. They say that his physics teacher once gave him a university textbook and sent him, along with a couple of classmates, smart like him, to a back room, telling them to learn on their own. Guth continued to excel at MIT, where he earned a bachelor, master, and Ph.D. in physics. He had a very promising future when he left Cambridge, MA, in the early 1970s, heading to a post-doc position at Princeton. However, towards the end of the decade, he found that he had

---

[4] On my website you will find a poetic description of the Big Bang: http://www.ice.csic.es/personal/elizalde/eli/inflacionariBB_eng_r.pdf.

already traveled the entire possible path of post-docs across the country: three years at Princeton, another three in Columbia, two more at Cornell, and one year at Stanford, without having yet obtained a tenure track. This meant being on the verge of absolute failure, despite the fact that these are among the best American universities. If you are not really hired in time, in the US you are out of the system and you must find some other way to make a living. Guth recounts that he asked a Cornell theorist to write him a recommendation for an assistant professorship, and he replied honestly that, at best, the letter he could write for him would be "*kind but vague*" (that is, ineffective). Corresponding to his multiple scientific interests, Guth had touched on too many subjects, and what he desperately needed at the time was a spectacular result in some particular line, whatever it might be (Fig. 6.19).

This actually happened in 1979. At the time, Guth was at the Stanford Linear Accelerator Center (SLAC), when he came up with a "spectacular realization" about the universe. He wrote this down in his notebook in capital letters with a box around it. Part of the problem had to do with his area of research: particle physics. Things happened this way. During those years Guth had spent a lot of time solving abstract mathematical problems related to the theory of elementary particles. However, a few months before, when he was at Cornell, a fellow post-doc, Henry Tye, had approached him with

Fig. 6.19 Alan Guth (1947–) at Trinity College, Cambridge, UK, in December, 2007. *Source* Betsy Devine aka Betsythedevine. CC BY-SA 3.0

a puzzling question about the early universe. Guth was intrigued and started trying to find an answer, even if that was not his field. "*I knew very little about cosmology,*" recalled Tye, "*but Alan, back then, knew even less.*" Some weeks later, Robert Dicke, from Princeton (whom we encountered above), went to Cornell to give a seminar. Tye arrived so late that the only available seats were in the back of the room, where he could barely hear Dicke. But Guth had arrived early, got a good seat, and took, as he always did, extensive notes. That night, at his home, he wrote an entry in his diary about the talk: "*fascinating*." Dicke had talked about the "*flatness problem*" in the Big Bang theory. As the universe has expanded for 14 billion years, tiny variations in the beginning should have been exaggerated by now, with a disastrous effect. Dicke had pointed out that, for our universe to look like it does today, one second after the Big Bang, the adjustment should have been within 15 decimal places, that is, in the interval between 0.999999999999999 and 1.000000000000001. However, the Big Bang theory offered absolutely no explanation of how this could have happened. It seemed crazy to suppose it was just a coincidence.

A second seminar, this time by Harvard physicist and future Nobel laureate Steven Weinberg, finally convinced Guth that this was a very important topic for reflection: the first infinitesimal fraction of the first second of the universe. Another question his friend Tye had asked him was about whether in the so-called grand unification theories there would be magnetic monopoles. Working together, Guth and Tye concluded that there should have been a vast number of magnetic monopoles in the early universe, and yet none are observed today. Guth was already at Stanford when he and Tye were finally trying to finalize their work in order to publish it. But they discovered in horror that another researcher, John Preskill, had done so ahead of them [132]. Driven by the need to publish, they proposed to complete the work by finding a solution to the problem of monopoles, which in the end they managed to do the following year [133]. Guth's mind was already peering into the precipice: he was going to be hopelessly out of work the following year.

According to Guth's own, very detailed account, inspiration came to him at around one o'clock in the morning on December 7, 1979, while his wife Susan and their two-year-old son Larry[5] were sleeping in the next room. In the morning, after riding faster than ever before on his bicycle to work, and after frantically doing some more numerical calculations related to the intuition he had had during the night, Guth wrote in his notebook, in large capital letters "*SPECTACULAR REALIZATION*," tracing a double box

---

[5] Larry and my eldest son, Sergi, took some graduate courses together while they were preparing for the Ph.D. in the MIT Department of Mathematics, at the dawn of the present millennium.

around those words. He would later name it cosmic inflation: an exponential expansion of the universe in an infinitesimal fraction of its first second. Next, he called Tye at Cornell and enthusiastically told him that this inflationary model solved not only their common monopole problem, but also the flatness problem Dicke had told them about—two very important problems, and simultaneously! Tye himself later acknowledged that: "*When Alan told me he had solved the flatness problem I had no idea what he was talking about.*" Guth asked him if he cared that he continued to work alone on this idea of 'inflation,' as he was in dire need of publishing, and Tye gave him permission to do so on the spot. Thus, Tye lost the greatest opportunity of his life (Fig. 6.20).

A few weeks later, over lunch at Stanford, Guth heard some colleagues talk about another cosmological problem he knew nothing about: the "*horizon problem,*" the inability of the Big Bang model to explain, with precision, the current homogeneity and uniformity of the temperature, in very distant places of the universe. Doing some more calculations in his notebook, he was as amazed as a child to discover that his inflationary model could also solve this problem, with as much elegance as the others. That was simply incredible. On January 23, 1980, just over a month after his 'night revelation', Guth made his model public at a seminar he gave at SLAC. Among the attendees was Sidney Coleman, a famous Harvard physicist who was on sabbatical there. At the end, he congratulated Guth in these terms "*every word you said was pure gold.*" Coleman immediately understood the extraordinary value of the proposal. Guth's solution involved a very short but very fast period of supercooling during a delayed phase transition, which produced a 'false vacuum' (an unstable state in time of the lowest possible energy density). As a result of the quantum tunneling effect, the false vacuum eventually decomposed into a true low-energy vacuum. And Guth had discovered that the decomposition of the false vacuum at the beginning of the universe could produce very surprising results, including an exponential expansion, now called cosmic inflation, which suddenly solved all the above problems.

And so it happened that, the next day, Guth already had a couple of job offers, and several invitations to take his seminar to the main campuses in the country. In the following months, job offers came from Harvard, Princeton, and half a dozen other top universities. But his dream was to return to his intellectual home, MIT, where, at that time, there were no job offers for the physics department. After giving his seminar tour and before accepting any of the numerous offers, but finally overcoming his shyness, Guth picked up the phone to call someone he knew at MIT, letting him know that if MIT could find him a position, he would accept it immediately. The next day, he

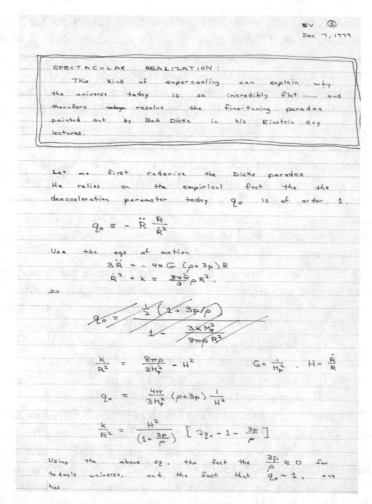

Fig. 6.20 Alan Guth's notebook with his first description of cosmic inflation. Guth's notebook is now part of a permanent exhibit at the Adler Planetarium and Astronomy Museum in Chicago. Fair use

got a call from MIT offering him, not just an assistant professorship, which was what Guth expected, but an associate professorship with an option to full professorship (which he later obtained, in 1986). His journey was over, although not his proposal for a theory, which he had to modify, and quite soon afterwards.

After accepting MIT's offer to join the faculty in the fall of 1980, he went on working on the details of his inflation model at Stanford [134]. In June, he discovered that a part of the model, namely the part relating to the horizon problem, was not working properly. Again, he felt a huge hole in his stomach, although he was relieved, as he later confessed, to think that "*at least I have already got a job at MIT!*" For some time, he could not find a way to get the universe out of the inflation stage—so that stars and galaxies would have the opportunity to be formed later—what is often-called the "graceful exit" problem. Indeed, he feared his theory could turn out to be a failure because of this. But, after reading in late 1981 a *preprint* [135] by the Russian physicist Andrei Linde (who had been working on the problem independently) and another one by Andreas Albrecht and Paul Steinhardt [136] on the same issue of the "graceful exit," Guth began to exchange *preprints*[6] with these colleagues. They helped each other until they finally arrived at a formulation of the theory of inflation. Since then there have been many other refinements and revisions based on Guth's original model and also introducing quite different ideas that lead to analogous results. Just listing them would take up a lot of space, so I shall not do that here.

Guth is still at MIT, where I have met him several times and attended his brilliant lectures and seminars. He has received many awards and distinctions, including the International Center for Theoretical Physics Medal and the Eddington Medal. His 1997 book, "*The Inflationary Universe: The Search for a New Theory of Cosmic Origins,*" became a best seller. An anecdote, which is well documented on the internet, is that in 2014 Guth was for a couple of months a very firm candidate for the Nobel Prize in Physics. He already had the aura of an NP winner, and there were even early champagne celebrations. The BICEP2 experiment located at the South Pole was believed to have found direct traces of cosmic inflation—the true Big Bang of the universe—in the polarization of the microwave background signal (CMB). I usually explain this in my talks, saying that Guth already had at this moment half a NP in his pocket, but only for a while. A reanalysis of the data found that the signal observed in the CMB was perfectly compatible with the polarization caused by the cosmic dust of the Magellanic Clouds region—towards which the instrument was pointing when making the observations—and which was, which was of the same order of magnitude. In application of Occam's razor principle, the result was rejected. Which does not mean that it was proven false, but this is how physics works: until it is checked again with perfect clarity, distinguishing the signal from the background noise, there is no case.

To complete this very interesting story, I recommend accessing a recent, very juicy colloquium: a podcast by Steven Strogatz in which he interviews

---

[6] See my essay E. Elizalde, *On how the Cyberspace arose to fulfill theoretical physicists' needs and eventually changed the World: Personal recalling and a practitioner's perspective,* in J. Martín Ramírez and Luis A. García-Segura Eds., *Cyberspace: Risks and Benefits for Society, Security and Development* (Springer Verlag, Berlin, 2017), pp. 3–22.

Brian Keating, the principal investigator of the BICEP2 project [199]. It is my opinion, shared by many highly qualified colleagues, that if this finding of the polarization traces in the CMB—corresponding to the primordial gravitational waves associated with inflation—had been confirmed, it would not only have been worthy of the Nobel Prize but, very likely, we would have found ourselves in front of the most important discovery of physics in the new millennium[7] In any case, I cannot fail to mention that there are also alternative approaches to the origin and evolution of the Universe, such as those corresponding to cyclic models [170, 200], to explain the initial stage, or those that use solutions other than Friedmann's (e.g., the black hole one) to describe its late and future evolution [182], to mention just a couple of ideas and references, from among the several possibilities that are on the market.

The reader may ask where this leaves the fundamental principle of energy conservation? According to Einstein's theory of relativity, the energy of the gravitational field is negative. The energy of matter, however, is positive. The entire universe creation scenario could unfold without breaking at any moment the law of energy conservation. The positive energy of all the matter in the universe may be precisely counteracted by the negative energy of all the gravity fields in the universe. That is not new: it is to repeat, once again, what we could already have read in Tolman's 1934 book and learned directly from Friedmann's equation.

However, it turns out that this is now much more than mere theory or possibility. Astronomical observations from the COBE, WMAP, and PLANCK satellites and others have been increasingly consistent with this idea. From the data, calculations of all matter and all gravity in the observable universe indicate that the two values compensate with amazing precision: all matter plus all gravity add up to zero. Consequently, the universe could come from nothing because it is fundamentally nothing, at least insofar as the energy balance is concerned (Fig. 6.21).

In the words of Guth himself, in one of his presentations:

It is rather fantastic to realize that the laws of physics can describe how everything was created in a random quantum fluctuation out of nothing, and how over the course of 15 billion years, matter could organize in such complex ways that we have human beings sitting here, talking, doing things intentionally.

The theory of inflation also predicts density perturbations, that is, tiny oscillations of the cosmos that serve as seeds for the subsequent formation of galaxies, as we now observe them. And they are just what inflation theory

---

[7] One of the authors of the BICEP2 paper was my former student Sergi R. Hildebrandt.

Fig. 6.21 President Barack Obama greets four of the nine 2014 Kavli Award recipients in the Oval Room of the White House, along with representatives of the Kavli Foundation and other guests, on July 31, 2014. The recipients were selected for their fundamental contributions to the theory of cosmic inflation, to the search for the limits of resolution of light microscopy, and to the discovery of specific brain networks for memory and cognition. Alan Guth is the one in the center, with his hands in his pockets. Official White House photo by Pete Souza (public domain)

suggests should have happened. Even so, things are obviously not so clear when today so many different forms of inflation have been proposed: double, triple, and hybrid inflation, modified hybrid, hyperextended, 'warm', 'soft', and 'tepid' inflation, 'natural' inflation, and so on and so forth. Inflation has ceased to be a theory and become a paradigm, or simply a generic idea that respects the essential features of Guth's original proposal. The very fact of finding acceleration in the expansion of the Universe, which we will discuss later, seems to confirm Guth's "miracle of physics No. 1": 'gravity' can indeed act repulsively (negative pressure, as we have already seen). In fact, the same kind of repulsive force that originally fueled inflation most likely also drives the current acceleration of the universe. Since mass and energy are equivalent, this repulsive energy must also gravitate, and if there were enough of it, it would preserve the original mass-gravity balance required to flatten the universe, which does seem to happen in fact, in accordance with the theory of cosmic inflation.

We can summarize by saying that, despite the fact that definitive evidence of inflation has not yet been obtained, the accumulated indications undoubtedly show us that we are on the right track. If not an inflation as exactly described by the various models of it, something quite similar to it must have happened, shortly after the beginning of time, so that the Universe can be as we see it today. But which is the correct inflationary model, or at least the best of those so far proposed?

6.6 R^2 or Starobinsky Inflation

Among the multiple theories of inflation, there is a second, very important one. Chronologically, it was even the first to appear, before Guth's, although it was not recognized as such until much later. I must stress that I will not mention here other proposals[8] that were made, some of them simultaneously with or even prior to Guth's work. None can compete (not even the one I am going to discuss now) in purpose, clarity, and definition of its goals with the one already considered.

I have spoken at length of the extraordinary beauty and usefulness of Einstein's GR and we have already verified that all modern cosmology is supported by this theory. However, we know perfectly well that this is not the *final* theory. It cannot be the ultimate one for the description of gravity, because it does not incorporate quantum effects. Quantifying gravity has proven to be an impossible task, at least until now; but not so trying to improve GR a little bit, by adding quantum *corrections*. The physicists of the Russian school, clustered around the great Yakov Zeldovich, were clearly the first to initiate such a program, in the 1970s. They worked on theories of particle creation by quantum effects in the early universe and around black holes. Hawking himself acknowledged that it was during a visit to the Soviet Union in the mid-1970s that he drew his inspiration to build his famous theory of quantum radiation from a black hole. Alexei Starobinsky,[9] himself a student of Zeldovich, observed that the aforementioned quantum corrections lead to second-order terms in the curvature, R^2. And in December 1979 he published a paper [137] on the spectrum of primordial gravitational radiation due to these quantum corrections, in the Russian journal *Pisma Zh.*

---

[8] And I fully accept criticism for this. My defence would be that this is not an exhaustive treatise on modern cosmology, but just a search for its roots; in particular, for those which, in my modest understanding, later developed and bore fruit.

[9] Starobinsky has been in the front row, where he likes to sit, in a good number of my presentations, over the years and across the globe, and I would like to thank him here for his comments, so often complimentary and always, without exception, constructive.

Eksp. Theor. Fiz. (which translates to *Journal of Experimental and Theoretical Physics Letters*). In another work [138], from 1980, he showed that such quantum corrections might even be able to eliminate the initial singularity of the universe, present in GR. It was not until years later, once Guth's theory of inflation was well known and accepted, that it was observed that theories with quantum corrections of the R^2 type (or with other higher-order terms in R) also gave rise to cosmic inflation, differing in several respects from that of Guth, but a kind of inflation nevertheless. And this is how, in the end, Starobinsky's work eventually became the first inflationary theory, since it had been published several months before Guth's (Fig. 6.22).

Starobinsky had no intention of solving the aforementioned problems with the Big Bang model. His aims were quite different. But, without intending or knowing it, he built the first inflationary model, before anyone else. However, it is not only that aspect, purely anecdotal if you like, that has made his model so important. It is precisely the physical background, so fundamental, that gave rise to his model: the terms added to GR are quantum corrections to classical gravity that must necessarily be there, at any rate. Terms of this

Fig. 6.22 Alexei Starobinsky (1948–) on 10 March 2014 at the Zeldovich monument in Minsk, Belarus, while participating in the Zeldovich-100 Conference. *Source* Melirius—DSC_9060. CC BY-SA 2.0

type appear, as expected, in the only fundamental theories, such as that of strings, which are capable (in principle) of quantifying gravity. That is why some specialists believe that R^2 inflation is the best and most 'natural' one.

7

Towards the Very Instant of Creation
of the Universe

We still have before us a couple of things pending. One is very important: to delve even deeper into the origins of the cosmos, right back to the very instant $t = 0$. Another is to complete the discussion on all the present meanings of '*Big Bang*', that is, all the different concepts that have adopted this same name. As I said, apart from the Big Bang model, which we have already considered, in the seventy years that have passed since Hoyle spoke these words, they have been used in very diverse situations. In particular, within physics, they have several other meanings, such as the Big Bang singularity, the Hot, Cold, and Warm Big Bang models, and already outside of physics, they have appeared in numerous contexts, in novels, movies, and in a very popular TV series. It is not surprising that the latter is currently the first meaning that inevitably appears when one searches the Internet. Therefore, each time the term "Big Bang" is used, one should readily specify which of these concepts is actually being referred to. Very often, such is not the case in the literature, and this adds some confusion.

So there several concepts that respond to the name "Big Bang." Today, in many scientific references, including drawings, models, and charts of the evolution of the Universe, the most common use of "Big Bang" is to refer to the singularity at the origin of the Universe. I always call this, very carefully, "the singularity of the Big Bang" or "the Big Bang singularity"; but I am pretty much alone in doing so. By saying "Big Bang," many are simply referring to the singularity, and not the model, nor the inflationary stage that occurred when the universe was already 10^{-33} or 10^{-32} s old (young?) and lasted

© The Author(s), under exclusive license to Springer Nature
Switzerland AG 2021
E. Elizalde, *The True Story of Modern Cosmology*,
https://doi.org/10.1007/978-3-030-80654-5_7

between 10^{-36} and 10^{-35} s. This would be much more appropriate, because according to Hoyle, "Big Bang" literally meant something like that when he defined this term with such clear words. But it is very difficult to fight against conventions and the evolution of the meaning of words with time.[1] As there is now another name to designate what Hoyle meant, e.g., inflation, the Big Bang has become "the mysterious origin," or "the initial singularity." It is hopeless to resist, but please try to be clear, from the beginning, on what we are talking about.

I will not go into all the possible meanings of "Big Bang" in great detail, but I will make a few points.

7.1 The Big Bang Cosmological Models

Just a brief mention, accompanied by a number of basic references, on this vast topic of cosmology. It was the center of heated discussions between several generations of cosmologists. Specifically, the controversy between the hot and the cold Big Bang models was finally decided, and very clearly, in favor of the first possibility. The question had a follow-up, or second part, that was even more important: with hot or with cold dark matter? That one was eventually decided in favor of the second possibility.

7.1.1 The Hot Big Bang Model

In the very remote past, but already after matter and energy had been created, the Universe was in a state of high density and extremely high temperature. Thus, its description could be reduced to that of a system with thermal equilibrium statistics: black-body radiation filled all space. It was a primordial soup of elementary constituents, namely quarks and gluons, the so-called quark-gluon plasma. With the expansion of the Universe, the temperature dropped until the first atoms, of hydrogen, were formed, and radiation was then able to travel, for the first time ever, through the entire Universe, giving rise to what we now call the cosmic microwave background or CMB radiation (formerly also called the cosmic background radiation or CBR). Some useful references where this process is described in detail are [139]. In the beautiful narration by Lemaître himself, the processes that subsequently took place can be compared with spectacular fireworks [8]:

---

[1] Suffice to observe the evolution of languages: over the years, the same Spanish word has sometimes come to have totally different meanings in the various countries of Latin America.

The evolution of the world can be compared to a firework display that has just ended; some few red wisps remain, ashes and smoke. Standing on a chilled cinder, we see the slow fading of the suns, and we try to recall the vanished brightness of the origin of the worlds.

In a time interval of between two and thirty minutes, but mainly within the first three minutes after the Big Bang (see [140] for a very popular reference), an efficient synthesis of the lighter elements took place, namely deuterium, helium-3, and helium-4. This is what is called the era of primordial nucleosynthesis. The current abundances of these light elements are in agreement with what happened under these conditions and during that time, and place strong restrictions on the state of the Universe, and particularly on its baryonic density. Our universe now contains around 23% of its mass in this primordial helium (its production in stars is not relevant, compared to the primary production during the first three minutes). The conditions then had to be precisely those that lead to our Universe, which has nine hydrogen nuclei for each helium nucleus [141]. Furthermore, it is now well known that most of the hydrogen in the Universe is in its simplest form and not in heavier isotopes, namely deuterium or tritium. Deuterium, in turn, is not produced, but only destroyed, in stars, so that its abundance today sets a lower limit on the amount of deuterium from primordial nucleosynthesis, and again also on the baryon density.

The Hot Big Bang model explains what we now see in our Universe. Summarizing the evidence, I limit myself to list what are usually called the four pillars of the standard Hot Big Bang model [142]:

1. The expansion of the universe.
2. The origin of the CMB.
3. The primordial nucleosynthesis of light elements.
4. The formation of galaxies and large-scale structures.

After six decades of using this model with considerable success, a crisis occurred in the early 1990s [143], which preceded the discovery of the acceleration of the expansion of the Universe and completely changed the whole paradigm [144], especially concerning the energy content of the universe. Some important consequences of this discovery, particularly regarding the possible appearance of future singularities in a finite time, will be discussed in the next section.

7.1.2 The Cold Big Bang and Other Models

The idea of a possibly cold Big Bang dates back to Lemaître's theory of the primitive atom, in the form of a large ball of nuclear liquid in a very low-temperature state. This was strictly necessary in order to maintain it without crumbling due to thermal fluctuations. In Lemaître's words [145]:

> When matter exists as a single atomic nucleus, there is no point in talking about space and time in connection with this atom. Space and time are statistical notions that apply to a set consisting of many individual elements; they were therefore meaningless notions, at the moment of the first disintegration of the primitive atom.

However, as advanced before, this idea could not explain the expansion of the Universe nor the origin of the light elements. A variant of Lemaître's cosmology was proposed in 1966 by David Layzer [146], who developed an alternative to the standard Big Bang cosmology, but which did not last long. He proposed that the initial state was close to absolute zero, which was reminiscent of Lemaître's initial state. From thermodynamic arguments, Layzer reasoned that, instead of starting in a state of high entropy, the universe began, on the contrary, with a very low entropy [147]. In any case, the evidence of the CMB radiation is very difficult to explain in this type of theory, although there have been some more recent attempts [148]. To conclude, most of the versions of the Cold Big Bang model that have been considered predicted an absence of acoustic peaks in the cosmic microwave background radiation. The WMAP data and, more recently the PLANCK data, definitively refute all these models. For completeness, other theories, which are somehow alternatives to the Hot Big Bang model, can be found here [149].

7.2 The Big Bang Singularity

7.2.1 The Belinsky-Khalatnikov-Lifshitz and Misner Singularities

In the 1960s, one of the main cosmological problems studied by Lev Landau's group in Moscow was the possible time singularity at the origin of the Universe. In particular:

(a) Whether cosmological models based on general relativity necessarily contain a temporal singularity; or

(b) Whether in fact the temporal singularity was a mere artifact of the hypotheses used to simplify these models (such as the homogeneity and isotropy of the universe).

In several articles, published between 1963 and 1971, Vladimir Belinsky, Isaac Khalatnikov, and Evgeny Lifshitz (BKL) demonstrated [150] that the Universe oscillates around a gravitational singularity, in which time and space become equal to zero. They also showed that this singularity was not artificially created by the simplifications introduced in obtaining solution families, such as the Friedmann-Lemaître-Robertson-Walker, quasi-isotropic, and Kasner ones. Their models were characterized by an anisotropic, homogeneous, and chaotic solution of the Einstein field equations of GR.

In 1969, Charles Misner built a similar model [151], called the mixmaster universe, which was also homogeneous, but not isotropic, and which expanded in some directions, while contracting in others. With the directions of expansion and contraction alternating repeatedly, which suggested that the evolution might in fact be chaotic.

7.2.2 The Classical Singularity Theorems (Penrose, Hawking)

Here I will summarize, in a unified way, only two of the main singularity theorems, among all those that appeared, in the mid-1960s, for Einstein's field equations free of any additional specific assumptions (such as homogeneity or isotropy). The starting point for all of them was Roger Penrose's pioneering singularity theorem [152] of 1965 (for some more references, see [153]). Penrose did in fact very brilliantly close the possible escape door discussed by the Landau school, by rigorously proving that, under very general assumptions, the singularity is inevitable. Penrose's proof is based on the concept of incomplete geodesics. His main results, avoiding the precise concepts (I refer the reader to the references given) are as follows. His first theorem refers to the conditions that occurred at the origin of the Universe, for which one needs only follow its evolution into the past, by running time backwards. So, under very general conditions, which stand true for our Universe (Hubble's law, essentially), and admitting the validity of Einstein's GR, our Universe had a beginning in a singularity, which everyone now calls *the Big Bang singularity*. This is checked technically by observing that all the temporal geodesics, extended to the past in time, are incomplete. Time itself ends: from there, there is no more time further back. And this happens, whatever path is taken to travel towards the past. There is no possibility of going further back. In Hawking's words: "*Time has its beginning in the singularity of the Big Bang.*" (Fig. 7.1).

The second theorem refers to black holes. It is very similar to the first one, since the conditions in the geometry of space–time that occur when one approaches a black hole are, in essence, the same as those observed when traveling back in time towards the origin of the Universe. In both cases, there exists what is called a *trapped surface*. Every observer that starts from it, even if it is a photon that can travel at the maximum speed possible, that of light, will inexorably find a future singularity, in which time disappears. There is

Fig. 7.1 Roger Penrose receiving his Nobel Prize medal and diploma at the Swedish ambassador's residence in London. ID: 110379. © Nobel Prize Outreach. *Photo* Fergus Kennedy, 2020. With permission

no beyond. Again, in Stephen Hawking's words: *"Time has its end in the singularity of every black hole."* (Fig. 7.2).

In short, time is born in the singularity of the Big Bang and dies in the singularity of every black hole. That is, in summary, the brief history of time. In any case, it is my duty to warn the reader against such categorical conclusions. These singularities, however rigorous and definitive they may seem, are *purely mathematical*, they do not have a firm physical meaning. The reader should now ask: Why? After all, I have spent a long time waxing lyrical about the beauty and extraordinary effectiveness of GR to describe the Universe in which we live and the absolute rigor and universality of the theorems I have just outlined. Yes, but let us see. By trying to reach the origin of everything or penetrate a black hole until hitting its very singularity, we have entered first quantum-physical domains and, further, regimes completely *unknown* to present day physics. Let us reflect a little. Even before intending to reach

Fig. 7.2 Black-and-white photograph of Stephen Hawking at NASA's StarChild Learning Center

these extreme limits (namely zero or infinity), if we try to apply the marvelous Einstein's equations to a single atom or molecule, we will miserably fail: they are not valid there! Luckily, for these levels we have the no less extraordinary quantum physics. But if we wish to go still further and get to $t = 0$ s or $x = 0$ cm, before doing this we have to get through, say $t = 10^{-50}$ s or $x = 10^{-50}$ cm. There, we already enter unknown territory: *no* physical theory has any answers there, not even quantum field theories. Without running too much risk, I would venture right now to bet a box of the best wine that even the quantization of gravity, if we had managed to build it, could not solve this problem.

In other words, as rigorous as the singularity theorems are—watertight from the *mathematical* viewpoint—their *physical* meaning is questionable, since the singularities they predict are in a region in which the equations that were used to get them no longer have the least physical validity. This is, of course, well known, and those were precisely the reasons why the Moscow school around Yakov Zeldovich (Starobinsky, Mukhanov, Chibishov, and others) sought to introduce quantum corrections into the equations of gravity. And that is precisely what, in practice, was verified by Starobinsky (as we have already seen), Hawking, and many others: simply by introducing an additional term as a quantum correction, it turns out that the Big Bang singularity may just disappear, and that black holes may no longer be completely black, but begin to emit radiation. It is not difficult to imagine how the whole landscape would change if gravity could be quantized. And, even more so, if it were possible to discover a valid physical theory beyond the Planck length, say, valid at 10^{-60} s. This is still absolutely unthinkable today.

7.2.3   A Brief Tribute to Stephen Hawking

As I write this, exactly three years have gone by since Stephen Hawking's death, and I cannot help paying a little tribute to him. My first encounter with Hawking was actually a virtual one. This would not be news now, quite the contrary, since today most if not all seminars and conferences given anywhere and at any level take place online. But I am speaking about the year 1975. I was then trying to finish my Ph.D. thesis, which I successfully defended the following year. And, as an advanced postgraduate student, my department at Barcelona University agreed to finance my participation in what was to be my first international school: the *XIV Internationale Universitätswochen für Kernphysik* [154]. It took place in Schladming, a beautiful town in the Austrian Alps, every couple of years always towards the end of winter. I got there after a long drive, which was wonderful in every way. With a couple of companions from Barcelona, we took our time, skirting the spectacular lakes of northern Italy and later going through the endless tunnels under the Brenner pass [155].

I could not have imagined the extraordinary two-week conference that was awaiting me; although, in fact, at that time I understood rather little of the lectures of such exceptional speakers. Julian Schwinger, a Nobel laureate in Physics, was the star. He had won the Nobel Prize in 1965, shared with Richard Feynmann and Sin-Itiro Tomonaga, for the quantization of electrodynamics. That was a real feat. In the words of Freeman Dyson, this theory had brought order and harmony to the domain of the intermediate fundamental forces of nature, leaving only gravitation and the nuclear forces out (at the time). Schwinger's way of attacking the problem was, however, very different from Feynmann's, who used his famous diagrams. Schwinger did this by means of analytical extensions, a tool of great mathematical elegance that I myself would later use for much of my life as a scientist in the context of the zeta function [156]. I remember Schwinger playing Austrian bowling with us, the younger ones, in the evenings. He shot very badly and seldom hit; however, he continued to play enthusiastically. But—and now comes the "virtual" meeting—all the lecturers, including Schwinger, were overshadowed during the conference by some results that Stephen Hawking had obtained, and that were brought to us by word of mouth. They had not been published yet, but everyone was talking about them: in the intervals outside the sessions, at lunchtime, and in the evenings.

Hawking was not present in Schladming, and his work was still a preliminary version, in *preprint* form (his article was later published in the prestigious journal Communications in Mathematical Physics [157]). I can assure the reader that the emotion that those results unleashed was highly contagious, even for me, who understood rather little about the subject. As I have already said, interest was growing with each passing day. I was very surprised by that situation, which I have always remembered.

It was only years later when I finally understood that all the interest and expectation that I had lived in Schladming were more than justified: for how was it possible for a black hole to emit anything? And I also learned later that some more veteran colleagues went to great lengths at first to prove that this could not happen. They tried hard to prove the falsity of that wonderful universal expression that Hawking had found for the temperature $T_{BH} = \hbar c^3/8\pi\, kGM$ at which every black hole radiates as if it were a perfect black body. A sublime, exquisite formula that combines in such a simple and precise way the most important fundamental constants of nature: Planck's constant $\hbar$ (of quantum mechanics), the speed of light c (of relativity theory),

Boltzmann's constant k (of thermodynamics), Newton's constant G (of classical physics), and the numbers 8π (representing mathematics and realizing the spherical nature of the hole); and ultimately making the temperature of the black hole, T_{BH}, only depend (inversely) on its mass M. Before long, however, even the fiercest critics had to admit that the formula was entirely correct (there are now half a dozen rigorous proofs, using very different approaches). And that, in addition, this expression opened an immense new field of research, until then almost unexplored, making a crucial connection between general relativity and quantum physics.

It is not surprising that Hawking's article now has a record number of about thirteen thousand citations, which is quite exceptional for a work with such a heavy mathematical content. One of the first to recognize, in all its significance, the extraordinary importance of Hawking's result was Paul Davies, who extended it to a different, but more common and general situation. Davies also worked with Hawking's preprint, as reflected in the list of references to his work [158]. A procedure analogous to the one used by Hawking to demonstrate the creation of particles by black holes was applied by Davies directly to the Rindler coordinate system for a flat spacetime. The result was that an observer subjected to a uniform acceleration κ would apparently see a fixed surface radiate at a temperature of $\kappa/2\pi$. In simple terms, and from our present perspective, it can be said that, while Hawking's finding opened the way to the concept of *black hole thermodynamics*, Davies greatly extended this notion to what was later called the *thermodynamics of space–time*. In another highly influential article, published a year later, William Unruh replaced the black hole with specific boundary conditions on the past horizon of space–time; and he showed that, when the detector was accelerated, even in flat spacetime, particles that would have been created by the quantum vacuum would be physically detected [159]. In this way, it was actually Unruh who established a very close connection between the two cases considered by Hawking and Davies. To be more specific, he rigorously established the close similarity of the case of an accelerated detector to the behavior of a detector near the black hole, demonstrating in particular that a geodesic detector near the event horizon would not see the flow of Hawking particles. In his article, Unruh reobtained Davies's analysis of the Hawking effect on flat spacetime with full mathematical rigor. We could thus say that he put the icing on the cake for Hawking's great discovery.

To end this point, I can't help but add two things. The first is that these authors, Davies in particular, recognized the influence that the ideas launched (although not published) by Bryce DeWitt in 1974 had had on their work, whose ultimate consequences are yet to be elucidated [153]. The thermodynamics of space–time, with a convenient formula for the entropy (such as Robert Wald's and its generalizations [160]), constitute today a sound basis for such important concepts as emergent and entropic gravity [161]. It has been proposed, in particular, that it could well be the case that gravity is not in fact a fundamental force, but a phenomenon that emerges from the thermodynamics of space–time itself [162]; and that it is perhaps closely related to the quantum entanglement of degrees of freedom, in a non-perturbative description of quantum gravity. But I am afraid I cannot go further along this path right now.

Hawking's main discoveries. On his website, Hawking defined himself as a *"cosmologist, space traveler, and hero."* Nothing more, nothing less. He worked all his life on the basic laws that govern the Universe. And he himself made it very clear, on his website, what his three main discoveries were.

1. With Roger Penrose, he showed that Einstein's general theory of relativity implies that space and time have their origin in the Big Bang and have their end in each of the black holes that form in the Universe. These are the so-called singularities, which necessarily appear in Einstein's theory. These results indicated that it was absolutely necessary to unify this theory with quantum physics, namely, the two great revolutions of physics of the first half of the twentieth century.
2. As a result of the search for such a unification, Hawking discovered that black holes cannot be completely black, but must emit radiation; and thus that they will finally evaporate and disappear. (This discovery is what caused such a sensation when I was in Schladming.)
3. And another conjecture, this time obtained in collaboration with James Hartle, is that the Universe in fact has neither edges nor boundaries, if time is imaginary. This would imply that the way the Universe began would be completely determined by the laws of science.

As can be seen, it is a curriculum vitae with just three items and less than half a page; as short as these discoveries are extraordinary. Here is another lesson that we should learn: to be suspicious of very long CVs with dozens of items, with which some try to leave us speechless. The more important the scientist, the shorter his or her CV.

Possibly the greatest of Hawking's discoveries is the second. In fact, I have already talked about it before. It is about the evaporation of black holes, also known as *Hawking radiation*, which he discovered in 1974:

Every Schwarzschild black hole, of mass M, emits electromagnetic radiation as if it were a black body at temperature: $T_{BH} = \hbar c^3 / 8\pi kGM$.

In his famous book *A Brief History of Time* (Bantam Books, 1988), Hawking wrote about this result: "*My work originated from a visit I made to Moscow in 1973, where the soviet scientists Yakov Zeldovich and Alexei Starobinsky suggested to me that, according to the uncertainty principle of quantum mechanics, rotating black holes should necessarily create and emit particles*". Between saying and fact, however, there is a long way, sometimes even an abyss. And this one Hawking was able to overcome, in that case, and in a very brilliant way, by giving a technically impeccable, beautiful, and quantitative demonstration of that prediction. It should also be added that his work masterfully complemented a previous result due to Jacob Bekenstein [163]:

Every black hole has an entropy, $S_{BH} = c^3 A / 4\hbar G$, which is finite and not zero. As must also be its corresponding temperature.

Here A is the area of the black hole's event horizon. Leaving aside the universal constants, this formula tells us that the entropy of a black hole is equal to a quarter of its area. This expression has also opened an entire new field in theoretical physics. The fact that the entropy of a black hole is proportional to the area and not to the volume of the same (as would be in the classical case) brings as consequence that quantum information can be measured from geometry in one dimension less, that is, in holographic terms. And this has given rise to very important developments in theoretical physics, which have already produced some brilliant results, connecting theories that did not seem to have anything to do with each other, in principle. Holography provides a non-trivial connection between field and particle theories of the conventional quantum physics and the proposals of quantum theories of gravity. As an example, an attempt is made to understand, with its use, superconductivity at high temperatures and some properties of superconductors, which are very difficult to study with conventional approaches.

Returning to Hawking's formula, among the direct consequences it has are the following:

- The Hawking radiation emitted by a black hole would be a perfect black body emission.
- Microscopic black holes would emit vastly more radiation and consequently disappear very quickly.

- If the extra-dimensional theories are correct, CERN's LHC laboratory could possibly create microscopic black holes. In fact, there are already projects trying to achieve this [164].

Just a short note about Hawking's last days. There are witnesses who assure that he was working until the end on what is known as the information loss paradox. However, I also observe that, in another article published posthumously [165], co-authored with his collaborator Thomas Hertog, the subject is quite different. This proves that Hawking was still capable of dealing with several problems at the same time, shortly before he died. In the work with Hertog, some theoretical characteristics of the inflationary Big Bang are re-examined through new mathematical proposals. The final version of the article appeared just ten days before Hawking's death (Fig. 7.3).

On the information paradox that obsessed him so much, we have the testimony of his collaborator Malcom Perry, who was at Harvard working on the preliminary version of an article on that, together with Andrew Strominger. Sasha Haco also collaborated in the work, in addition to Hawking. Perry still did not know that Hawking was so seriously ill, and he called him with the sole intention of personally updating him on the progress he and Strominger had been making lately. Undoubtedly, this was the last scientific exchange that

Fig. 7.3 Laying the urn with Hawking's ashes in his grave at Westminster Abbey. Copyright Dean and Chapter of Westminster, London. Reproduced with permission

Hawking had with his colleagues. According to Perry's testimony: "*It was very difficult for Stephen to communicate and I was put on a loudspeaker to explain where we had got to. When I explained it, he simply produced an enormous smile. I told him we'd got somewhere. He knew the final result.*"

In October 2018 his colleagues Malcolm Perry, Andrew Strominger, and Sasha Haco published what was to be his last article [166]: "*Black hole entropy and soft hair,*" which examines the so-called information paradox of black holes, a puzzle to which Hawking dedicated a good part of his life. The problem is briefly that, despite the fact that he himself had discovered that black holes radiate and thus, in principle and with enough time, all the matter that they have previously swallowed could be recovered, the same does not happen with the information carried by the engulfed matter. This is because it turns out that the Hawking radiation that is emitted is perfect blackbody, so does not carry *any* information. This is thus completely lost behind the black hole's event horizon. As it happens, this contradicts quantum physics, which has among its fundamental principles the principle of unitarity, which entails the conservation of information. Here we face a really puzzling paradox (Fig. 7.4).

The solution proposed by the four authors in the above-mentioned article is broadly speaking the following (the technical aspects are all but easy to explain). If the black hole had some "hair" just above the horizon, that "hair" could actually retain (filter, if you like) the information associated with the matter that falls on the hole, just before it disappears through the event horizon. In this way, the information would not be lost! It would remain tangled in the hair, thus retained on the outside of the event horizon. This work was completed, as I said, in the days before Hawking's death, and the published article also includes a tribute to all of his important contributions.

Malcolm Perry adds, speaking about that work: "*It's a step on the way, but it is definitely not the entire answer. We have slightly fewer puzzles than we had before, but there are definitely some perplexing issues left.*"

Unfortunately, or perhaps fortunately—it all depends on how you look at it—this turns out to be very often the case. I mean, some important open questions in science remain unsolved when the great geniuses die, even though these have dedicated their entire lives and all their energy to trying to answer them, and despite the fact that, in the best of cases, they have actually managed to answer a few of these crucial questions along the way.

Here are some important problems that still remain open, in this context:

• The paradox (also called catastrophe) of the information loss, as black holes evaporate.

Fig. 7.4 Stephen Hawking's gravestone in the nave of Westminster Abbey, London, where his ashes are interred, not far from the graves of Sir Isaac Newton and Charles Darwin. The inscription is an English translation of a phrase which appears in Latin on Newton's gravestone. The stone depicts a series of rings, surrounding a darker central ellipse. The ten characters of Hawking's equation express his idea that black holes in the universe are not entirely black but emit a glow, that would become known as Hawking radiation. In this equation, T stands for temperature, $\hbar$ stands for Planck's constant, which is used to understand quantum mechanics, c stands for the speed of light, 8π is related to the spherical nature of the black hole, G is Newton's gravitational constant, M stands for the mass of the black hole, and k stands for Boltzmann's constant, which is the energy of gas particles. The Caithness slate stone was designed and made by John Maine and the letter cutter was Gillian Forbes. Copyright Dean and Chapter of Westminster, London. Reproduced with permission

- The problem of the initial conditions of the universe.
- The Big Bang singularity.

Hawking's discoveries were remembered at a memorial service held in June 2018 at Westminster Abbey. In the course of it, the DSA2 antenna in Cebreros (Spain), oriented towards the near black hole 1A 0620–00, broadcast his voice, wrapped in Vangelis' music, for a few minutes. A gift for other intelligences that may be inhabiting our cosmos. What Hawking achieved, the great discoveries he left us as a scientist—like the beautiful formula with which I started this note and which is now etched on his tombstone in the nave of the Abbey, not far from those of Newton and Darwin—will remain forever. So, too, will the memory of his extraordinary example as a person.

7.2.4 On Borde-Guth-Vilenkin's Theorem

Without going as far as I had proposed above, inflationary cosmological models already seemed to invalidate the conditions of all classical singularity theorems. For this reason, in the 1980s, attempts were made (although without success) to build models that, based on de Sitter's exact solutions, could be eternal. In 1994, Arvind Borde and Alexander Vilenkin proved an extended theorem [167], which states that, even with inflation, space–time remains geodesically incomplete towards the past, which implies as before the existence of an initial singularity. The hypothesis for this theorem was that the energy–momentum tensor should obey the weak energy condition (WEC). That was already a step forward with respect to the previous theorems.

However, quantum corrections to inflationary models do violate this condition when quantum fluctuations cause an increase in the Hubble parameter: $dH/dt > 0$; on the other hand, this is an essential condition for chaotic inflation to be eternal. Therefore, the weak energy condition had to be generically violated in those models. Again, this seemed to open the door to escape the conclusions of Borde and Vilenkin's theorem, and was the motivation of Arvind Borde, Alan Guth, and Alexander Vilenkin in their celebrated article [168] "*Inflationary space-times are incomplete in past directions*". As the title already indicates, they reverted back to the earlier result of Borde and Vilenkin, but with some important additional considerations which had not been fully appreciated by certain staunch supporters of creationist theories. I am not going to discuss this issue here though [169]. Technically, one now begins from the (sufficient) condition of a "quasi-de Sitter" state with a minimalist hypothesis about "averaged expansion" along temporal paths: the Hubble constant averaged over all time trajectories must be positive, $H_{av} > 0$. Furthermore, the theorem can be extended to additional dimensions, and also to cyclical models [170].

As a final summary, in all these cases, and under the strict conditions of the theorem, geodesic incompleteness continues to be obtained. In other words, under seemingly reasonable conditions, time has an origin even for inflationary models with quantum corrections. Let me leave it here.

7.3   How Could the Universe Have Originated?

The most widespread view today is that the origin of the Universe took place from nothing (better, from almost-nothing; for a more detailed description see, for example [171]; and for some critical comments, see [172]). If we think a bit about this, the first obvious question is: But, what is *nothing?* The answer of current day physics is (at the very least) double:

1. **In classical physics**, the fundamental theory is GR, and there the *vacuum solution* is the de Sitter solution, i.e., the zero-energy solution of Einstein's field equations.
2. **In quantum physics**, on the other hand, "nothing" translates as the *vacuum state* of the quantum system under consideration; in our case, the one at the very beginning of everything, as far as we can go into the past with the best fundamental physics that we know today.
3. It is quite obvious that we are missing a new theory here, which we would need in order to answer this question properly, namely a theory of *quantum gravity* (QG).
4. However, it should be added that it is not at all clear that, even in possession of QG, we would be allowed to penetrate the *Planck domain*, which sets a limit for all known theories of physics—and very probably also for this still unknown QG.

As of today, this is the state of the art regarding this fundamental question. Leaving aside the Planckian constraint and other considerations of rigor, some possibilities have recently been proposed. We can consider two minimalistic starting points:

1. **Only quantum space-time**, and nothing else. Although there have been some attempts to achieve this (such as one a few years ago by Lawrence Krauss and Frank Wilczek [173]), no one has yet been convincingly successful, and the opinion prevails that this is not actually possible.
2. In addition to space–time, **a scalar field (or two)**, namely the Higgs, an inflation field, perhaps others. Naturally, this already seems a more feasible possibility, at the expense of having to explain where those additional fields come from.

To complete everything I have said, there is a wonderful table in the following link which I highly recommend to readers, detailing one after the other all the different stages of the evolution of our Universe: https://en.wik ipedia.org/wiki/Chronology_of_the_universe.

So, how do current physical theories conceive of the very moment of the creation of the Universe? For Roger Penrose and Stephen Hawking, that instant is a mathematical singularity, remaining out of reach of any physical interpretation. But it turns out that quantum corrections can radically change such a result. And there are new models (Alex Vilenkin, Andrei Linde, and others have been working on them for more than twenty years) in which they combine inflation and quantum fluctuations of the empty state of a primitive system, within which a spark (a scalar field, a Hawking-Turok instanton, etc.) would be able—at zero energy cost—to initiate an inflation process that would dramatically amplify the tiny fluctuations (of order the of the Planck scale) of the quantum vacuum. These are present, it should be remembered, due to Heisenberg's uncertainty principle (one of the pillars of quantum physics). In order to give rise to the fluctuations that we clearly observe in the cosmic microwave background (CMB). These give us the photograph of the oldest map of the Universe that we have so far; it dates back to when it was only about 378,000 years old We will see this later.

But let us go back to the origin and start again. When the cosmic inflationary expansion stopped, almost all of that colossal energy produced in the space transformed into the elementary components of ordinary matter and energy: quarks, gluons, leptons, photons, and so on, which filled the still very small universe, although part of this energy was used to heat them (what we now call reheating). All these elemental constituents formed an enormously hot primordial soup, which is called the primordial plasma (or the quark-gluon plasma). It was a thoroughly dark universe, since photons could not travel through it: when one managed to leave a material particle, it could not take two steps without being trapped by another that would absorb it. And so on, and so on, and so on. Absolute darkness reigned; there was no light.

And all that plasma was beating in unison, as if it were a kind of universal heart (the beats are called baryon acoustic oscillations, BAO). The changes that took place in the plasma, its evolution from the first few seconds to the first few minutes of the universe, are wonderfully described in books such as Weinberg's and also in Wikipedia, as already mentioned above.

Exactly the same conditions of the primordial plasma that made up the universe—when it was only a few thousandths of a trillionth (10^{-15}) of a second old—have been reproduced with enormous precision, on a small scale, in particle physics laboratories, such as the LHC of CERN, in Geneva. When the cosmos was already one hundred thousandth of a second old, protons and neutrons were formed, and when it reached the end of its first second of life, neutrinos were suddenly uncoupled from the plasma and could travel through it. Not so photons. Not yet.

Attempts are currently being made to obtain information about this stage of the universe from primordial neutrinos, which are just beginning to be detected in certain experiments. During the time from ten seconds to three minutes, approximately, protons and neutrons came together to form the lightest atomic nuclei (hydrogen, deuterium, helium, etc.), in a process called primordial nucleosynthesis. However, it took much longer until the atoms could be formed. Much of the theory of particle physics is involved in the study of these processes, which I have just sketched rather roughly here. This is indeed a fascinating field of study.

The universe continued to expand at an already normal rate, much like the one we now detect. And, for this reason, it became less and less hot, since the expansion itself was gradually cooling it. Then, when it was some 378 thousand years old, an episode occurred that is masterfully described in the first verses of the Book of Genesis: suddenly, *there was light!* The temperature had dropped until it was below the ionization threshold of the smallest atom, hydrogen. All at once, these atoms precipitated out on a large scale and the dark primordial plasma, the entire universe, suddenly became transparent to photons; these could travel for the first time from end to end of the still very young cosmos. The light of the first cosmic dawn invaded the entire Universe.[2]

This was the very first light in the universe, a wonderful homogeneous and isotropic glow of black body radiation that we have been able to observe with the help of the curious 'eyes' of the COBE, WMAP, and Planck satellites. This is what we call the cosmic microwave background (CMB), which captures our imagination without fail. Let me repeat: this is the first light of the cosmos, the light from the first day of Genesis; a glow that will never be turned off and continues to travel throughout the current universe. A light which, in addition, bears in it the indelible traces of the last beat of the primal plasma. And, also, the stretched traces of the quantum fluctuations of the pre-inflationary era. And the traces, in short, of all the events that have occurred in past times in the history of the universe, which influenced the CMB while it was traveling toward us; and that we are increasingly capable of deciphering, from our standpoint, on our small planet (a blue dot) somewhere in this immense universe. Tell me, now, if what we have learned about the origins and whereabouts of our cosmos is not wonderful, a miracle, a marvel.

---

[2] On my website you will find a poetic description of the CMB: https://www.ice.csic.es/personal/eli zalde/eli/The%20First%20Light%20of%20the%20Cosmos_1.pdf.

We have certainly come a long way from Lemaître's model of the primordial atom or cosmic egg. Now, in the same way that we could have asked him where the hen capable of laying such an egg could have come from, you could equally ask me where the pre-primordial space–time foam come from, and the pinhead of initial matter, the Higgs field, the inflaton, and all the rest. But let us be clear: it is one thing to put the entire *jivarized* (shrunk) universe in our top hat, and quite another to just hide a few tiny elements in it, which we would never be able to see even with the most powerful microscopes imaginable. My top hat is much emptier than the one any magician will show you. And from them, from these very tiny elements, and with a "simple" blowing up of the space that makes up the universe's balloon (with time being its radius), we have been able to create a truly enormous universe and the whole of its material and energy content. And what's more, at zero energy cost, in accordance with what I already explained above.

My description has been very sketchy, but it contains the essentials of everything we know today. I have not mentioned the important later stages, such as recombination and the formation of the first galaxies and primordial black holes, and nor have I more than touched upon the subject of stellar nucleosynthesis, which Hoyle pioneered and which gave rise to the heavy elements, as I mentioned before. Many questions remain open, but this is science: we should never expect the absolute, definitive truth from it. Whoever seeks this must look to other sources. Note also that the above description is already on the furthest frontier of the fundamental physics that we know of. It cannot yet be said that cosmic inflation has actually been proven. But there are many important indications that it did indeed happen. And what the alternative theories (of a pulsating universe, in loop cosmology, etc.) try to do is essentially to recreate the same effects, although starting from other principles. Anyway, what is clear, what we have checked beyond any doubt, is that there is not the slightest trace of the primaeval atom or the great explosion that many misinformed fellows continue to refer to; we can be 100% sure that this never happened.

Another observation, this one for advanced readers, is that I did not mention the multiverse at any point. Actually, the possibility is not excluded that the whole universe could be infinite and that everything I have described so far just happened in a very small domain of a huge multidimensional space–time; that only a tiny portion of the same inflated, etc. In other words, my description could just refer to our universe. It may thus be that there are a huge number of other universes like ours being created and disappearing all the time. Superstring and brane theories do admit such possibilities. But there have been no real physical proofs so far. We still have a long way to go before we can answer the questions I have raised. We do not have a theory that unifies quantum physics with gravity. But I have already mentioned elsewhere that it is quite possible that this much desired theory, even if it were found, might not be enough to reach the zero-point, $t = 0$, the Big Bang singularity, which unfailingly appears in our theories of gravity—and for which, among other things, Roger Penrose has just been awarded the 2020 Nobel Prize in Physics. Far from discouraging anybody, this should push us to continue the investigation.

Finally, with all the astronomical observations, the so-called standard cosmological model, or ΛCDM (Cold Dark Matter model with a cosmological constant, Λ) has been firmly established, as I will later illustrate. However, in order to go further—eventually to $t = 0$—we will need better 'eyes', namely detectors capable of capturing and systematically processing better data (LIGO, LISA, BBO, DECIGO, etc.), and maybe at some point the traces of primordial neutrinos and primordial gravitational waves (PLANCK, the BICEP Array, the Simons Observatory, LiteBIRD, etc.), which we hope to be able to process within a decade or two. With this, photographs of a much younger Universe will be obtained and, predictably, inflation will be confirmed.

One last but not less important thought. It is healthy, from time to time, to look back and see where we have come from and everything we have managed to understand so far; how we are actually progressing, step by step, year by year, century by century. We should not get obsessed, once again, by those questions that are still very difficult to answer, and which (as our experience of past discoveries has shown) it may take a hundred years, or even more, for science to answer.

And we can be completely sure that, by then, new questions will have arisen in search of an answer.

8

The Second Cosmological Revolution of the Twentieth Century

Astronomical observations carried out, just over twenty years ago, by two teams of some thirty scientists each (Riess et al., Perlmutter et al., 1998)[1] indicate that the expansion of the Universe is taking place at an accelerated rate. The first group to publish results was the *High-z Supernova Research Team*, led by Brian Schmidt and Adam Riess, in 1998. The other, known as the *Supernova Cosmology Project*, with Saul Perlmutter as principal investigator, did so independently, the following year [174] (Fig. 8.1).

According to most of the sources easily available, this discovery was an extraordinarily surprising one, and fully unexpected. What both teams were actually looking for, was just to verify the strength with which the distribution of masses of the Universe slow its expansion speed, due to the gravitational attraction. They just intended to corroborate previous results they had found in that direction. But it happened that both groups, independently and with mutually compatible results (although, for some time, they had serious discussions about that, rarely recalled now), found, to their amazement, that the cosmic expansion continued apace, thereby overcoming the gravitational attraction. However, this is not the full story, to say the least. For real specialists in the subject these results were not surprising at all, quite on the contrary, since there were already many hints of a non-vanishing cosmological constant implying that the universe's expansion was accelerating (Fig. 8.2).

---

[1] The well-known researcher María Pilar Ruiz Lapuente, from the University of Barcelona and formerly an undergrad student of the author of this book, was a member of the second collaboration.

© The Author(s), under exclusive license to Springer Nature Switzerland AG 2021
E. Elizalde, *The True Story of Modern Cosmology*,
https://doi.org/10.1007/978-3-030-80654-5_8

a   b   c

Fig. 8.1 **a** Adam Riess at the Nobel Prize 2011 press conference at the Royal Swedish Academy of Sciences. *Source* Holger Motzkau, Wikipedia/Wikimedia Commons. CC BY-SA 3.0. **b** Saul Perlmutter at the Nobel Prize 2011 press conference at the Royal Swedish Academy of Sciences. *Source* Holger Motzkau, Wikipedia/Wikimedia Commons. CC BY-SA 3.0. **c** Brian Schmidt at the 62nd Lindau Nobel Laureates meeting on July 5, 2012. *Source* Markus Pössel (User name: Mapos)—Own work. CC BY-SA 3.0

In fact, to start with, the idea that the universe contains close to homogeneous dark energy that approximates a time-variable cosmological "constant" had arisen in particle physics many years before, through the discussion of phase transitions in the early universe and through the search for a dynamical cancellation of the vacuum energy density; and similarly, in cosmology, through considerations of how to reconcile a cosmologically flat universe with the small mass density indicated by the peculiar velocities of galaxies. Moreover, in both fields, it was clear that Λ had to be at present quite small, since it would have been rolling toward zero for a long time already.

In addition, from the mid-1980s the cold dark matter theory with a nonvanishing cosmological constant was beginning to be seriously considered as a much better fitting model to different astronomical data than the ordinary CDM model (with no Λ) [175]. Early discussions of its significance for large-scale structure had appeared, as [176, 177], while the anisotropy of the 3 K cosmic microwave background temperature in the ΛCDM model had been analyzed in 1985 already [178, 179]. This is just to name a few of the many papers that seriously considered the possibility of an accelerating universe (for more extensive details, see [180, 181]). One can thus say that the really surprising outcome of the SNIa analysis would had been that these results had *not* confirmed the cosmic acceleration! We see here one more example of a crucial point misleadingly explained in many places.

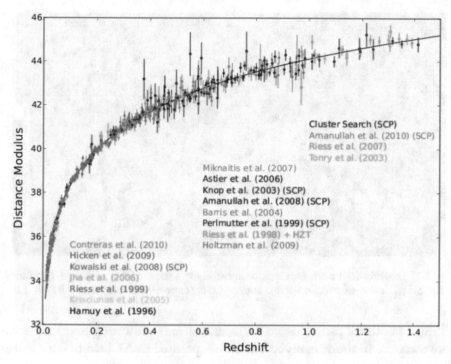

Fig. 8.2 The expansion of the universe is not slowed down by gravity, but, on the contrary, accelerates, as the parabolic behavior of the fitting curve clearly shows. Suzuki et al. (The Supernova Cosmology Project), Astroph. J. 746, 85 (2012), arXiv: 1105.3470 [astro-ph.CO]. The SCP "Union2.1" SNIa compilation is an update of the "Union2" compilation, now bringing together data for 833 SNe, drawn from 19 datasets. Of these, 580 SNe pass usability cuts. We present new data from the HST Cluster Survey. Copyright Supernova Cosmology Project. Fair use

Further to that issue, the ΛCDM model was not only on the minds of many cosmologists, but it had actually been used for detailed simulations in the late 1980s and had already been employed as a very effective tool to compare with observations of structure formation and clustering. When my former student Enrique Gaztañaga was a postdoc at Oxford in 1993, the whole group there used the ΛCDM simulations of George Efstathiou et al. and Enrique himself published several papers in 1993–1995 comparing ΛCDM with observations and clustering data, power spectrum, and higher order correlations. We also had some papers before those years where we started to use a non-vanishing Λ. In addition, there were all these results for the cosmic background radiation in the search for the first acoustic peak, to which the two of us contributed together with our student Pablo Fosalba,

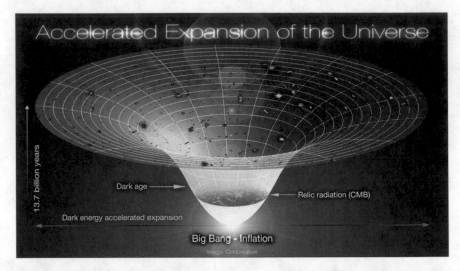

Fig. 8.3 Lambda-cold dark matter, accelerated expansion of the universe, Big Bang-inflation (timeline of the universe). *Source* Coldcreation—own work. CC BY-SA 3.0

all them published much before SNIa, and already indicating that the curvature was zero. In short, many results clearly pointed to ΛCDM being a better fit to astronomical observations.

The ΛCDM model subsequently became the leading one, following the observations of the accelerating expansion in 1998, but *only* after having been supported by other thoroughly independent and very important observations, among them the detection of the integrated Sachs-Wolfe effect by Gaztañaga, Fosalba and collaborators [182]. In particular, in the year 2000, the BOOMERanG microwave background experiment measured the total (matter–energy) density to be close to 100% of the critical one, whereas in 2001 the 2dFGRS galaxy redshift survey measured the matter density to be only some 25%. The large difference between these values strongly supported a positive Λ or dark energy, with a relatively important contribution, actually the dominant one. Much more precise spaceborne measurements of the microwave background from WMAP in 2003–2010 and Planck in 2013–2015 have continued to support the model and pin down the parameter values with great accuracy, below 1% uncertainty in most cases. They will be summarized later (Fig. 8.3).

Observations in support of the cosmic acceleration came later from many and varied directions and data analysis, including: the CMB, early galaxy formation, type 1a supernovae, nearby galaxy distributions; BAO via 21 cm hydrogen emission, gravitational lensing, B-mode polarization of the CMB signal and the amplitude of matter fluctuations the important parameter σ_8,

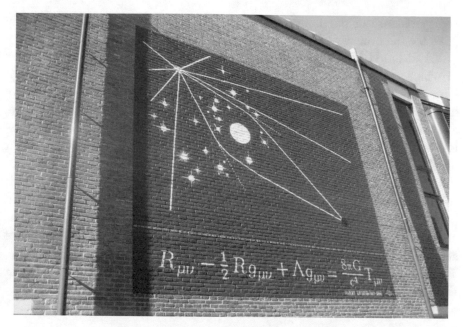

Fig. 8.4 The cosmological constant, Λ, was introduced by Einstein in 1917, to counterbalance the attractive force of gravity. Einstein (gravitational) field equations including the cosmological constant, Λ, by Albert Einstein, from his theory of general relativity (Stephen Weinberg, Gravitation and cosmology, New York, 1972 p. 155); painting by Jan-Willem Bruins (TegenBeeld); photograph by Vysotsky. Diagram of gravitational lensing (deflection of light by an intervening mass) with formula of Einstein on a wall of Museum Boerhaave, Leiden (the Netherlands). The line sketched is in fact a Morse code. Date: 13 July 2016. CC BY-SA 4.0

which measures the possibility that the universe could have formed stars and quasars early enough, and should be larger than 0.8 or so (the weighted average of WMAP and Planck data points is now positioned at $\sigma_8 = 0.8163$), Sunyaev-Zeldovich measurements, and Lyman-α forest results (Fig. 8.4).

Anyway, after this extensive explanation of the second revolution in cosmology that occurred towards the end of the last century the big question still remains: what is the *physical* reason for this universal acceleration? For that to take place, as we know from Galileo, 16th C., and Newton, 17th C., there must be a force that is repulsive and stronger than the gravitational attraction, and this unknown force must act on a cosmic level. But, what is the nature of such an everywhere present repulsive force that is able of causing the acceleration of the very 'fabric' of the cosmos? As in the case of dark matter, we have not yet been able to give an answer to this crucial question, more mysterious, if possible, than the one related to dark matter, and to which, furthermore, it corresponds a significantly greater fraction of the total energy balance of our Universe (some 72%, as compared to approximately 23% for the other).

Various models and experiments have been proposed for this,[2] the simplest of which fully respects the validity of Einstein's general theory of relativity, with the simple addition of his famous cosmological constant which, as we saw, he introduced to stabilize the static universe model prevailing in his time, and which would now be able to propel its accelerated expansion. The physical nature of this constant was a mystery to Einstein, but we now know that it could be a direct manifestation of the quantum fluctuations of the vacuum state of the physical fields that inhabit the universe [183]. Although difficult to understand, we could accept at first glance that this answer is beautiful and perfectly natural, were it not for the following amazing fact: when calculations are carried out in detail, the numbers do not match, giving a discrepancy of astronomical order (and never better put!), of many orders of magnitude. There is, therefore, a serious disagreement between theory and observation at this point. This is known as "*the cosmological constant problem*".

8.1   An Expanding Universe Satisfies Einstein's Equations

Reflecting for a moment, as soon as we know a little bit of physics, we can easily understand that a universe in uniform expansion, like the original Big Bang model, does not need the action of any force to continue in this way, expanding indefinitely. An initial boost is sufficient for it to happen, although only as long as the density of matter/energy in the universe is less than a certain critical value, since otherwise a high density would be able to slow down the expansion of the universe until it reduced it to zero, from which point it would begin to contract until it ended in what is known as a *Big Crunch* (Fig. 8.5).

Let us try to be more pedagogical. Imagine a child, the *Little Prince* of Saint-Exupéry's tale, say, who lives on a tiny planet of quite small radius. If the Little Prince threw away a stone with all his strength, it might never fall back down on top of him, but instead be lost in space, moving away forever. However, if the Little Prince tried to throw the stone away from Earth, we can be sure it would end up falling back on him. I am talking about the same physical principle here: gravitational attraction. Once again, a Universe with

---

[2] https://www.darkenergysurvey.org/; http://sumire.ipmu.jp/en/; https://twitter.com/ACT_Pol; http://hetdex.org/hetdex/; http://www.sdss3.org/surveys/boss.php; http://sci.esa.int/euclid/.

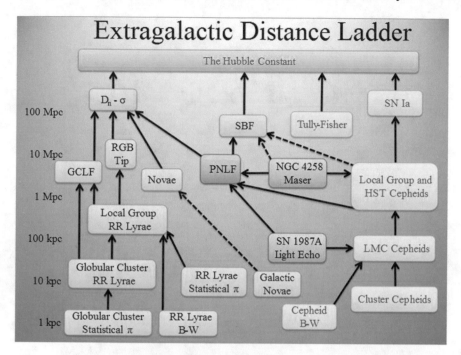

Fig. 8.5 Methods used to measure distances in astronomy, at different scales. Light green boxes: technique applicable to star-forming galaxies. Light blue boxes: technique applicable to population II galaxies. Light purple boxes: geometric distance technique. Light red box: the planetary nebula luminosity function technique is applicable to all populations of the Virgo Supercluster. Solid black lines: well-calibrated ladder step. Dashed black lines: uncertain calibration ladder step. *Source* brews_ohare—This file was derived from: Extragalactic Distance Ladder.svg. Created: 20 February 2009. CC BY-SA 3.0

a matter density less than some critical value would never be able to stop its expansion completely: it would continue for ever and ever (Fig. 8.6).

Until just twenty years ago, before the discovery of accelerated expansion, the crucial, most fundamental question of all cosmology was to find out whether the aforementioned matter/energy density of the universe was actually above or below this critical value. In other words, whether the universe would continue expanding forever or if, on the contrary, this expansion would stop completely at a future time, and from there, begin to contract until the Big Crunch.

Fig. 8.6 If the Little Prince threw a stone with all his strength, from his small planet, it would not return. Description: The Little Prince. *Source* http://www5e.big lobe.ne.jp/~p_prince/le_petit_prince_club_prive/img/book/The%20Little%20Prince% 206th.jpg. Painted in 1943. Author: Antoine de Saint-Exupéry (1900–1944)

8.2   But it Turns out that the Expansion of the Universe Is Accelerating

The paradigm has completely changed with the discovery that the expansion of the universe is accelerating. This finding is one of the greatest landmarks in the entire history of humanity. This makes dark energy, as I have already said, intrinsically much more mysterious, fundamental, and interesting than dark matter. Where does the force come from that constantly acts on the stone that our Little Prince has thrown and continues to accelerate it incessantly? Put another way, what is the origin and nature of the cosmic force that constantly accelerates the expansion of the Universe? In fact, this acceleration has not yet been proven to be constant. For this, we would need to demonstrate that its time derivative is zero, which is not possible with the quality of the data we have today, even the most accurate among them.

On a purely mathematical level, and with the appropriate sign, Einstein's cosmological constant is well capable of performing such a function. When he proposed it, in 1917, there was no physical explanation for this term. But at present we have known, for a number of decades that the physics obeyed by the bowels of the Universe (its atomic and subatomic levels) is quantum. And we know this now with incredible accuracy, reaching one part in 10^{18}, i.e., *eighteen* decimal places. All the phenomena that took place in the Universe when it was very small and hot were governed by quantum physics, which states in particular that, in any imaginable system, even its lowest energy state, the so-called empty state, "*nothingness*", in other words, is forever disturbed by by quantum fluctuations which, despite being in principle virtual, can physically manifest themselves as a real force that is measurable to very great accuracy in the laboratory (the Casimir effect and others) [9] (Fig. 8.7).

This force is not a weird entelechy, or something difficult to observe, but quite the opposite: it is an omnipresent force, common to all quantum systems and which, in principle, must always be taken into account, although it is sometimes very small compared to the other forces that act in our system (electromagnetic, or even gravitational) and it is usually completely negligible, since it cannot compete with them. Everything depends on various factors that must be analyzed in each specific situation. At the cosmic level the fluctuations of the quantum vacuum should, according to the most common theories, give rise to a much greater force than that observed, for example, in the acceleration of the universe. As I have advanced before, the values do not match with each other and there is no generally accepted explanation of this highly annoying fact. It is perhaps the greatest challenge in fundamental physics today, and therefore also in cosmology. Some great physicists, including distinguished Nobel laureates, such as Steven Weinberg [184], have been working on this question for many years. In an act of despair, they have even brought in the (not so scientifically orthodox) anthropic principle in a last ditch attempt to obtain an explanation for the small value of the cosmological constant and also for the extremely strange fact that the effects of quantum fluctuations are not observed at all at large scales (Fig. 8.8).

Fig. 8.7 Where does the force come from that constantly acts on our Little Prince and that continues to accelerate him incessantly, defeating the gravitational attraction of his planet? The Little Prince © Wikimedia Commons

8.3   Modifications to the General Theory of Relativity

If this path were shown to have no way out, we still have another possibility, which is to modify Einstein's equations, or what amounts to the same, to modify the general theory relativity itself (at least on large scales). Then, one enters into the realm of theories like $f(R)$ [185] and scalar-tensor theories. This possibility does not arise as a mere patch to cover the hole that has suddenly occurred in the magnificent construction of GR, at very small scales or, what comes to the same, at very high energies. I have already outlined the crucial idea before: quantum corrections to GR are actually unavoidable; they *must* be considered. This requires us to include additional quantum terms,

a   b

Fig. 8.8 **a** Yakov Zeldovich (1914–1987), cropped from an image of a Russian postage stamp. Stamp issuing authority—MARKA Publishing & Trading Centre. Printer—Association GOZNAK of the Ministry of Finance of the Russian Federation. **b** Steven Weinberg (1933–) at the 2010 Texas Book Festival, Austin, Texas, USA. *Source* Larry D. Moore. CC BY-SA 3.0

and then powers of the curvature, R^n, naturally appear. Now, since we are not sure that these perturbing terms are all that would be dictated by a (still to be constructed) quantum theory of gravity, it is not unreasonable to consider other functions $f(R)$, with some physical criteria, such as the fact that these terms do show up in the most fundamental theories, like string theories or loop quantum cosmology. Moreover, we now see more indications of the necessity for these modifications of Einstein's general relativity, which would make them more realistic and more far-reaching theories of gravity. This is so because such theories would be able to cover the whole history of the universe, from the very low scales of its beginning up to the current epoch and extending into the future. My group and collaborators have decisively contributed to such developments (Cognola et al. [186], Elizalde et al. [187]). Finally, my last consideration is mathematical (and rather non-trivial, actually). It turns out that the introduction into the theory of these new terms in the curvature is equivalent, from a mathematical perspective, to leaving Einstein's GR as it was initially, without touching a hair of it, but introducing additional fields, of scalar and tensor type. The simplest such models consist in considering the cosmological constant as a function of time, as a field, while others consider the added fields as cosmic fluids that permeate the universe.

And with this we have just reached the current frontier of our knowledge in theoretical cosmology and fundamental physics. In view of the way things are at present, as I finish this book, I foresee that we may have to wait a few years, or even decades, for the outcome of this story. I summarize below some of the scenarios and questions currently being researched.

8.3.1 Acceleration: New Singularities

In many of the modified gravity models that have been studied to obtain the accelerated expansion of the Universe, new singularities have appeared. Some of them are similar to those of the Big Bang and black holes, but others are very different, and occur in a future time. Starting from the second Friedmann equation, which we gave in an earlier chapter, and adding to it the equation of state corresponding to a possible cosmic fluid: $p = \omega \, \rho$, where p is the pressure, ρ the density, and ω the so-called equation of state parameter, the following three possibilities appear, depending on the different values of this fundamental parameter ω.

a. The most recent and precise astronomical observations indicate that the value of ω is very close to -1, which corresponds precisely to the fact that the current universe is dominated by Einstein's cosmological constant term. This has a very small value, given by the results of the astronomical observations already mentioned, and other more precise ones that were carried out later by several groups.

b. The main dichotomy concerns whether the value of ω is in fact a little greater than -1, in which case the fluid that would permeate the universe (a kind of cosmological constant, but variable in time) has been called *quintessence*; or whether the value of ω is slightly less than -1, which is what the data seem to favor today, within some margin of error. In the latter case, we enter into unknown physics, with the so-called *phantom energy*, which leads us to a series of *future singularities* in a finite (or infinite) time from now.

The singularities appearing in $f(R)$ models have been classified (see, e.g., Nojiri et al. [188]). However, quantum gravity effects may remove such singularities as well as, presumably, the Big Bang singularity. This was demonstrated, among others, by the author and collaborators in the seminal paper Elizalde et al. [189].

Not only in the last, but in all the cases above, many questions arise that are still unanswered. I will only mention two, corresponding to the simplest possible situation: *Why is the value of the cosmological constant the one it is?* Namely, the one imprinted in the sky, which we read from the astronomical data. This value has varied over time so, *why at the present moment does it have a value of the same order of magnitude as that corresponding to all the other energies of the universe, dictating the evolution of the cosmos?*

And in this context of the new theories, we must inevitably speak once again of James Peebles, Physics Nobel Prize awardee of 2019, whom we saw, together with Robert Dicke and David Wilkinson, as one of the authors of the theoretical prediction of the CMB. Another among his many important contributions was the work he carried out with Bharat Ratra in 1988. They were pioneers and forerunners of a great line of research in current theoretical cosmology, which hatched ten years later with the extraordinary discovery, mentioned above, that the expansion of the universe is accelerating. To repeat, such acceleration is very difficult to explain in physical terms and its contribution, called dark energy, to the total energy balance of the current universe is enormous, being more than 70%. What is remarkable is that Peebles and Ratra were so many years ahead when they published their premonitory works in which they analyzed the consequences of introducing a scalar field at the cosmological level, to reconcile the low dynamic estimates of the mean mass density with the spatial curvature, insignificantly small, preferred by inflation. This allowed them to have a model ready, well in advance, to develop the idea that space contains energy whose gravitational effect approximates that of Einstein's cosmological constant, thus anticipating by more than ten years the concept of dynamic dark energy, particularly in its form known as *quintessence* [190]. His work was essential in establishing a solid theoretical base on which to place all these revolutionary astronomical observations (Fig. 8.9).

Peebles has got many important awards. The announcement when he received the Shaw Prize reads: "*He has laid the foundation for almost all modern research in cosmology, both theoretical and observational, transforming a highly speculative field into a precision science.*" In addition to making major contributions to Big Bang nucleosynthesis, dark matter, and dark energy, Peebles was the primary pioneer of the theory of cosmic structure formation in the 1970s. He contributed decisively to laying the foundations of current physical cosmology and did a lot to turn it finally into a modern science [191]. In his own words [192] (Fig. 8.10):

It was not a single step, a critical discovery that suddenly made cosmology relevant; the field gradually emerged through a series of experimental observations. Clearly, one of the most important during my career was the detection of the cosmic microwave background (CMB) radiation that immediately caught the attention of both experimentalists interested in measuring the properties of this radiation and theorists; all joined to analyze the implications of the discovery.

Fig. 8.9   James Peebles (1935–). *Source* Juan Diego Soler. Created: 17 May 2010. CC BY 2.0

I cannot fail to mention here a personal anecdote. When the first three-dimensional map of the universe appeared in the mid-1980s (well, just a slice of it, in fact, Fig. 1.10) [193], this had a spectacular impact and was the origin of my increasing dedication to theoretical cosmology.[3] However, nothing would have been the same without the books by Peebles [194]. They

---

[3] Although the slice of the famous Harvard CfA scan made by Valérie de Lapparent, Margaret Geller and John Huchra (Fig. 1.10) contained only 1100 galaxies, the most important thing was that for 584 of them it was possible to determine their distance from us (through their cosmological redshift). For the first time in history, this allowed us to see a section of the Universe in three dimensions. The repercussions of this work were spectacular, also due to the structures that appeared in this point distribution: one could see a human form (the man), while another seemed to be a finger pointing toward us (God's finger), but the most intriguing feature was those huge empty areas, without any galaxy, surrounded by dots, that drew the above-mentioned structures. Many theoretical physicists from around the world, and astronomers who had not previously devoted themselves to large-scale cosmology, set to work at once, the former trying to create models that would explain the formation of these structures from fundamental theories of physics, the latter trying to find new observational confirmations of these behaviors of galaxies on a large scale. Sometimes they collaborated with each other, as was the case with Edward Witten and Jeremiah Ostriker. In Spain, the present author was among the first to be captivated by that map and immediately started working on the subject with his then doctoral student in search of a thesis subject (and now a reputed cosmologist) Enrique Gaztañaga. This was the germ that would give rise, over the years, to several teams of great impact in theoretical and observational studies of the Universe on a large scale, which definitely contribute to the pride today of Spanish science.

Fig. 8.10 James Peebles delivering the speech at the Nobel banquet, 10 December 2019. ID: 99755. © Nobel Media AB. *Photo* A. Mahmoud. Reproduced with permission

were the holy grail of my Ph.D. students.[4] I still remember the excitement with which the third one [195] was received, when it appeared in 1993. I will just add that in our group from the Institute of Space Sciences (ICE-CSIC) and from the Institut d'Estudis Espacials de Catalunya (IEEC), in Barcelona, Spain, which I lead together with Prof. Sergei Odintsov, we have been working for years on this type of models, with collaborators from all over the world. And I think it is fair to add, especially for the benefit of my wonderful collaborators, that together we have carried out sound and serious work that has had a high international impact in our field.

8.4 What Is Our Universe Like, According to Astronomical Observations?

Our Universe is now in the era that is named ΛCDM (i.e., cold dark matter with a cosmological constant Λ), and will probably remain in that same era in the future to eventually become, asymptotically in time, a de Sitter universe.

---

[4] Enrique Gaztañaga met him personally later, at the request of Peebles, who invited him to lunch to discuss his model, and they were about to collaborate, what actually did not happen (at least formally).

But if the data were to twist a little, by fine-tuning the error margins, and it turned out that we were actually in one of the other two possibilities, namely that of phantom dark energy or that of quintessence, other singularities would then be expected to appear [196] (Fig. 8.11).

Let us look briefly at the astronomical data, to confirm what I have said. They clearly indicate that the current observable universe corresponds to a geometrically flat space–time, containing a mass/energy density equivalent to 9.9×10^{-30} g/cm^3. The primary constituents consist of 68–72% dark energy, 23–26% cold dark matter, and 4.5–5% of atoms. This density of atoms is equivalent to a single hydrogen nucleus for every four cubic meters of volume [197]. The exact nature of dark energy and cold dark matter remains a complete mystery to this day. We have set these wide margins because the values based on astronomical observations continue to fluctuate as of now. I include a graph to illustrate this fact, where the results of WMAP are compared with those obtained by PLANCK. It would be absurd to attempt to go further when not even the true specialists agree, owing, as I say, to the error margins, which are still somewhat large (Fig. 8.12).

I add other important values to the above, corresponding to the curvature of the universe. These clearly indicate that its total energy is practically zero, as had been suspected for a number of years. I take the latest results corresponding to the cosmological legacy of the Planck satellite, in

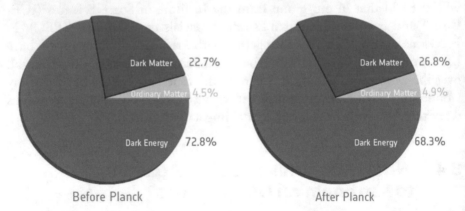

Fig. 8.11 Energy content of the universe according to WMAP and PLANCK. Planck's high-precision cosmic microwave background map has allowed scientists to extract the most refined values yet for the universe's ingredients. Normal matter that makes up stars and galaxies contributes just 4.9% of the universe's mass/energy inventory. Dark matter, which is detected indirectly by its gravitational influence on nearby matter, occupies 26.8%, while dark energy, a mysterious force thought to be responsible for accelerating the expansion of the universe, accounts for 68.3%. The 'before Planck' figure is based on the WMAP nine-year data release presented by Hinshaw et al. (2013). ESA and the Planck Collaboration. Space Science. Planck. Fair use

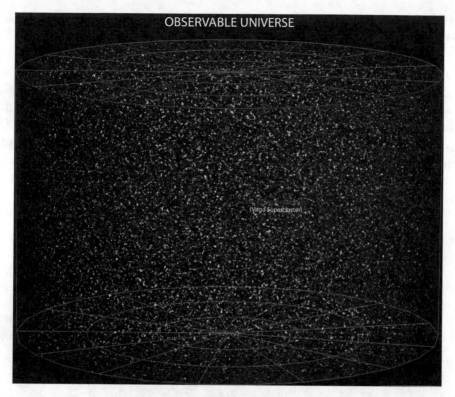

Fig. 8.12 The observable universe. The full compilation of the image is located here: Earth's_Location_in_the_Universe_(JPEG).jpg. Author: Anrew Z. Colvin, 2011. CC BY-SA 3.0

N. Aghanim et al., *Planck Collaboration, Planck 2018 results (The cosmological legacy of Planck)*, Astronomy & Astrophysics 641, A1 (2020); https://www.aanda.org/articles/aa/full_html/2020/09/aa33880-18/aa33880-18.html, and N. Aghanim et al., *Planck Collaboration, Planck 2018 results (VI. Cosmological parameters)*, Astronomy & Astrophysics 641, A6 (2020); https://www.aanda.org/articles/aa/full_html/2020/09/aa33910-18/aa33910-18.html.

Spatial curvature. Combining Planck data with BAO (Planck TT, TE, EE + low E + lensing + BAO):

$$\Omega_K = 0.0007 \pm 0.0019 \, (68\%)$$

The joint results suggest that our Universe is spatially flat to a 1σ accuracy of 0.2%.

Hubble constant. The *Planck* base-ΛCDM cosmology requires a Hubble constant

$$H_0 = (67.4 \pm 0.5)\,\mathrm{km\,s^{-1}\,Mpc^{-1}},$$

in substantial 4.4σ tension with the latest local determination by Riess et al. (2019) [198]. The *Planck* measurement is in excellent agreement with independent inverse-distance-ladder measurements using BAO, supernovae, and element abundance results. None of the extended models that the *Planck* collaboration has studied in the paper convincingly resolves the tension with the Riess et al. (2019) value of H_0 (Fig. 8.13).

Fig. 8.13 View of NGC 6357, a diffuse nebula in the constellation Scorpio. The nebula contains many stars in formation, submerged in dark disks of gas and dust, and other young ones that show twisted gas cocoons. Composite image produced by data from four telescopes: Chandra X-ray Observatory, ROSAT Telescope, Spitzer Space Telescope, and UK Infrared Telescope. *Source* NASA, 2016

Number of families of quarks and leptons. Allowing for extra relativistic degrees of freedom, the *Planck* collaboration has measured the effective number of degrees of freedom in non-photon radiation density to be

$$N_{eff} = 2.89 \pm 0.19 \, (N_{eff} = 2.99 \pm 0.17 \text{ including BAO data}),$$

which is consistent with the value 3.046 obtained in the standard model.

Parameter of the equation of state of dark energy. Combining *Planck* data with *Pantheon* supernovae and BAO data, the equation of state of dark energy is tightly constrained to

$$w_0 = -1.03 \pm 0.03,$$

which is consistent with a cosmological constant (but still with a small deviation in favor of the *phantom* case). The *Planck* collaboration has also investigated a variety of modified-gravity models, finding no significant evidence for deviations from ΛCDM.

Sum of the neutrino masses. Allowing for a free degenerate active neutrino mass, and combining with BAO measurements, the *Planck* collaboration obtains a tight 95% constraint on the sum of the masses

$$\sum_{\nu} m_\nu < 0.12 \, \text{eV}.$$

8.4.1 Some Conclusions from *Planck*

To end this chapter, although what follows is written mainly for professionals and some concepts may be difficult to grasp, I find it interesting to recall a few of the closing paragraphs of the conclusions of the *Planck* legacy paper and the one on the cosmological parameters, from September 2020 (see above). This is where we were one year ago.

"This is the final Planck collaboration paper on cosmological parameters and presents our best estimates of parameters defining the base-ΛCDM cosmology and a wide range of extended models. As in PCP13 and PCP15 we find that the base-ΛCDM model provides a remarkably good fit to the Planck power spectra and lensing measurements, with no compelling evidence to favour any of the extended models considered in this paper. Compared to PCP15 the main changes in this

analysis come from improvements in the Planck polarization analysis, both at low and high multipoles.

The overall picture from Planck, since our first results were presented in PCP13, is one of remarkable consistency with the 6-parameter ΛCDM cosmology. This consistency is strengthened with the addition of the polarization spectra presented in this paper. Nevertheless, there are a number of curious "tensions," both internal to the Planck data and with some external data sets. Some of these tensions may reflect small systematic errors in the Planck data (though we have not found any evidence for errors that could significantly change our results) and/or systematic errors in external data. However, none of these, with the exception of the discrepancy with direct measurements of H_0, is significant at more than the 2–3σ level. Such relatively modest discrepancies generate interest, in part, because of the high precision of the Planck data set. We could, therefore, disregard these tensions and conclude that the 6-parameter ΛCDM model provides an astonishingly accurate description of the Universe from times prior to 380,000 years after the Big Bang, defining the last-scattering surface observed via the CMB, to the present day at an age of 13.8 billion years (Fig. 8.14).

The flatness of the spatial hypersurfaces has been established at the 5×10^{-3} level. Neutrino masses have been constrained to be (0.1 eV). The number of relativistic species is consistent with three light neutrinos and disfavours any light, thermal relics that froze out after the QCD phase transition. The baryon density inferred from the acoustic oscillations up to $t = 400,000$ yr is consistent with that inferred from BBN at $t = 3$ min. Dark-matter annihilations are tightly constrained. Dark energy is consistent with being a cosmological constant that dominates only recently. The pattern of acoustic oscillations in temperature and polarization power spectra implies an early-Universe origin for the fluctuations, as in the inflationary framework. The primordial fluctuations are Gaussian to an exceptional degree. There are no gravitational waves at the 5% level, suggesting the energy scale of an inflationary epoch was below the Planck scale.

The ability of the ΛCDM model to explain the Planck data, and a wealth of other astrophysical observations, indicates that our understanding of physics is good enough to model 14 Gyr of cosmic history and explain observations out to the edges of the observable Universe. However, the surprising ingredients required by this model suggest that our understanding is highly incomplete in several areas.

Despite these successes, some puzzling tensions and open question remain. While many measures of the matter perturbations at low redshift are in excellent agreement with the predictions of ΛCDM fit to the Planck data, this is not true of all of them. In particular, measurements of the fluctuation amplitude from cosmic shear tend to lie low compared to the Planck predictions. Measures of the distance scale from nearby Type Ia SNe remain discordant with the inferences

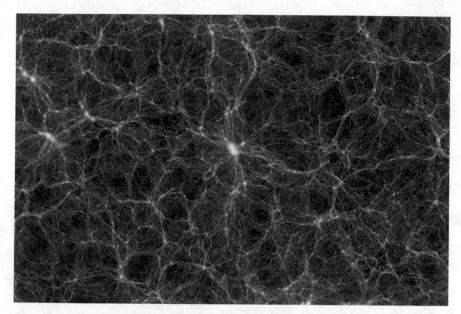

Fig. 8.14 The cosmic web. A vast, mysterious structure that links up far-flung galaxies, was observed directly for the first time in 2019. The observations reveal that an ancient cluster of galaxies about 12 billion light years away in the constellation of Aquarius are linked together by a network of faint gas filaments. The existence of the cosmic web is central to current theories of how galaxies first formed following the big bang, but until now evidence for it had remained largely circumstantial. User: Manasa Shastry. CC BY-SA 4.0

from the inverse distance ladder. We expect these areas will see continued atten-tion from the community, which will determine whether these tensions point to statistical fluctuations, misestimated systematic uncertainties, or physics beyond ΛCDM (Fig. 8.15).

Nevertheless, it is important to bear in mind that the main ingredients of ΛCDM, namely inflation, dark energy, and dark matter are not understood at any fundamental level. There is, therefore, a natural tendency to speculate that "tensions" may be hints of new physics, especially given that the landscape of possible new physics is immense. In the post-Planck era, the CMB provides enor-mous potential for further discovery via high-sensitivity ground-based polarization experiments and possibly a fourth-generation CMB satellite. The next decade will see an ambitious programme of large BAO and weak lensing surveys, and new techniques such as deep 21-cm surveys and gravitational wave experiments. Uncovering evidence for new physics is therefore a realistic possibility. What we have learned, and the legacy from Planck, is that any signatures of new physics in the CMB must be small."

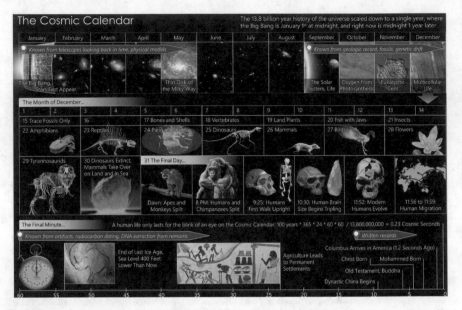

Fig. 8.15 A graphical view of the Cosmic Calendar, featuring the months of the year, days of December, and the final minute. The 13.8 billion year lifetime of the universe mapped onto a single year. At this scale the Big Bang takes place at the instant of midnight going into January 1, and the current time is the end of December 31 at midnight, and the longest human life is about 1/4th of a second, a blink of an eye. The scale was popularized by Carl Sagan in his book The Dragons of Eden and on the television series Cosmos, which he hosted. Author: Efbrazil, 2011. Original work of Eric Fisk. Images are composited from Wikipedia, NASA, and other government sites. CC BY-SA 3.0

9

Conclusion

I still owe the reader a couple of explanations. The first, to briefly summarize what Stigler's eponymy law, mentioned above, is about. Stephen Stigler, professor of statistics at the University of Chicago, stated it forty years ago [94]. It affirms, categorically, the following:

No scientific discovery is named after its discoverer.

Someone may immediately say that this is not true, that he or she knows cases in which this simply does not stand. This is true, but it is a proven fact, when analyzed in depth and considering a sufficiently large data set, that the cases that contradict the law are indeed few; they are just exceptions to it. Let us discuss some examples:

- As we have already seen at the beginning of this book, the Pythagorean theorem was known to Babylonian mathematicians a whole millennium before Pythagoras stated it.
- Halley's Comet had already been observed by astronomers since at least 240 BC.
- Coulomb's law was discovered by Henry Cavendish.
- The l'Hôpital rule was developed by Johann Bernoulli.
- The Oort cloud was postulated and described by Ernst Öpik.
- The all-famous Hubble law was obtained and explained by Georges Lemaître two years before Edwin Hubble obtained it.

© The Author(s), under exclusive license to Springer Nature
Switzerland AG 2021
E. Elizalde, *The True Story of Modern Cosmology*,
https://doi.org/10.1007/978-3-030-80654-5_9

- The Fermi Golden Rule was first used by Paul Dirac.
- And even Stigler's own law is no exception, since Stigler recognizes sociologist Robert K. Merton as its first discoverer, and even though Stigler himself tried to get it known as Merton's law, he has not succeeded.

When I presented my arguments in favor of adding Slipher's to the naming of the Hubble-Lemaître law, several colleagues, including the staunchest advocates of my point of view,[1] very wisely warned me about the futility of spending my time on such an undertaking, commendable as it might be. They largely supported their arguments by appealing to Stigler's law: "*We all know already that no discovery is named after its original discoverer, so why waste time?*" I understand and accept that argument willingly. But even then, it is one thing to simply label a discovery with a name (after all, there must be some label in the end); but quite another one to continue teaching in our classrooms and conferences (if only to simplify and not go into more details, due to lack of time or whatever) that 'the name appearing in the label' did everything, without even mentioning any other contribution. It is the latter that, in my opinion, is intolerable and dishonest. Especially now that—after spending so much time in reviewing old papers and journals to tie up loose ends—we finally know the truth. Or, at least, we have come much closer to it.

With this, I have exposed again my motivation in writing this book. It can be blamed on me, and quite rightly, that I have forgotten, along the way, some relevant facts and many other actors (even very important ones) who also participated, with their results, in the construction of the eventually accepted theories. I fully accept the criticism, my only defense being the limited scope of this book, apart from my own ignorance, which may have shown through on too many occasions. Let me say, however, in my defense, that this book claims only to be a guide towards finding the true origins of some basic concepts, after having selected certain crucial moments on which I place more emphasis. It is in no way a comprehensive compendium of modern cosmology. And I encourage my colleagues that may read my work to complement it, filling in the important gaps that surely remain.

Another point I want to highlight is that, throughout this search, a few remarkable heroes have emerged, most of whom, if not all, have not been able to conceal important weaknesses, however. That is so, in fact, because life itself abhors absolute perfection. But I am afraid I may have given the

---

[1] Some of them, who have held in their hands Slipher's writings—many of which are unpublished and kept in the archives of the Lowell Observatory—wrote to me that I was indeed even more right than I thought.

wrong impression that I was gloating when recounting the weaknesses of the great scientists. Nothing could be further from the truth. It is worth warning those who do not know this, that in science (and I guess in other branches of knowledge, too), what counts in the end, what remains for history, is *only* what was done well. With just a simple formula, many have entered the book of history forever. Here there are no final balances, as in accounting, where successes add up and failures subtract, in the computation at the end of each financial period. In science one only adds up everything that one has discovered, everything that one has done well, clearing the way for others to follow. All the rest, however serious one's error, is not subtracted in any way; it does *not* count negatively. It is simply forgotten. Only extraordinary achievements count, and for good. Hubble is no less Hubble because he was not able to understand the expansion of the universe, nor for having engaged in torpedoing the attempts of others to understand that the universe was expanding. The same is true of Einstein, who is no less of a genius no matter how hard he resisted, for ten years of his life, the physical reality that others so aptly showed him. This is the fate of scientists.

I keep the last point to highlight the final conclusion I have come to after this long journey. It is, as I already announced before, that the knowledge acquired in the last century gave rise to two important cosmological revolutions. Such a statement may seem, at first glance, exaggerated, given that, no matter hard we search through the literature, we are not going to find anywhere (to my knowledge) explicit reference to such revolutions. On the other hand, all encyclopedias speak of the Copernican revolution, and Galileo's and Newton's groundbreaking contributions. In addition, we have just seen what Peebles himself thinks about this: "*it was not a single step, a critical discovery that suddenly made cosmology relevant, but the field gradually emerged through a series of experimental observations.*"

All this is true, and I subscribe to it without hesitation; but what I say does not contradict Peebles in any way. I also admit that it is not possible to determine a fundamental treatise or a spectacular observation that has played the role of foundational fact or critical discovery (in the way that the taking of the Bastille, or the publication of "*The Origin of Species*" were). The development of modern cosmology was at all times a collective work like few others, in which many dedicated actors have taken part, both astronomers and theorists, with many great successes and many gross errors, too. The most I have managed to identify, for reasons that I am not going to repeat, are some dates, or rather, some significant time intervals, which I summarize thus:

1. **The first revolution** can be framed in the period from 1912 to 1932, that is, from the astronomical discoveries of Leavitt and Slipher to those of Hubble, and it includes the theoretical advances made by Einstein, Friedmann, de Sitter, and Lemaître. We may say that it came to its peak in 1929, with the publication of Hubble's results. Eventually, the scientific theory of the expansion of the universe having an origin was adopted by the main specialists and rounded off in the celebrated Einstein-de Sitter model of 1932.

   Anyway, it still had to wait for the elaborate formulation of the Big Bang model, for its definitive test through the detection of the cosmic background radiation (CMB), and for a major and crucial reshaping (inflation), which would only arrive fifty years later and was actually the prelude of the second revolution.

2. **The second revolution** occurred in the period from 1985 to 2005. Indeed, it may be affirmed that it started in the mid 1980s already, with the first discussions of a fitting cold dark matter model endowed with a possibly non-vanishing Λ-term; it came to its peak in 1998–99 when the supernovae results appeared, but it was not fully confirmed until some six years later or so, by other very important, supplementary surveys, as we have seen.

   To repeat, the ΛCDM theory that was being built around various cosmological results had its most impacting confirmation through the astronomical discovery of the accelerated expansion rate using Type Ia supernovae as standard candles. However, these results just confirmed what other astronomical observations and the theoretical fits had been stubbornly indicating for some time already: that the universal expansion is accelerating! Modified gravity models and the use of the cosmological constant provide solid theoretical frameworks to try to understand this astonishing fact, based on the fundamental laws of physics. But so far no one has convincingly succeeded.

The similarities between the two revolutions are very remarkable. Their gestation period was exactly the same: twenty years in both cases. In addition, there were always some prominent leaders who pretended to be the actual discoverers and tried to claim all the merit of the breakthroughs exclusively for themselves (Hubble and his Mount Wilson observatory, in the first case, Riess, Perlmutter et al. and their SNIa observations, in the second). What we have here discovered is that those claims are not sustained. The actual story was always much richer and more complex (as Peebles himself remarks), and involved many other very crucial contributions along the way leading first to the findings and then to the final confirmation of the discoveries.

It is also true, moreover, that the cosmological revolutions I am talking about cannot be identified by looking only at what happened during those 20-year time intervals. I have to admit that this may be just a sketchy approach, a possibly too crude simplification. Cosmological models are continuously being reshaped, as more and better astronomical data are being accumulated. But one thing remains clear. It is then and only then, when we now look in perspective at the entire past century, and observe it in this perspective view from its end to its beginning, strictly comparing the vision of our Universe at the end of the twentieth century with that at the beginning of, it is namely this view, from the outside and with a wide angle that finally opens our eyes to an incontestable reality. No one can deny that the detection of the CMB was in fact, without a doubt, the final confirmation of a revolutionary discovery, the tangible proof that our universe had an origin some 13.8 billion years ago and that it expands—in absolute contrast to the model of the visible universe that had reigned before, viz., small, static, eternal, and unchanged for ever and ever.[2] And the same can be said of the supernovae results, in pointing out the universal acceleration, confirmed afterwards.

The cosmological revolutions are unlike the great revolutions, so well documented, of physics (the theory of relativity and quantum mechanics) and mathematics (associated with Gödel's incompleteness theorem), in which, in response to observations of nature or through the consistency of the construction of pure mathematics, respectively, theories of enormous scope and depth have been created, breaking through previous ways of understanding things. In the case of cosmology, in the revolutions that I have identified, the fundamental contribution came from the astronomical observations of the cosmos, far more than from any theory. But it should not be forgotten that they took full advantage of the above-mentioned revolutions of physics, and in particular Einstein's general relativity, which allows us to interpret and understand the results of all those observations. No more but no less than that!

In short, the astronomical observations that I have referred to, which occurred gradually throughout the century, had extraordinary, almost incredible implications, which completely changed our vision of the Universe. From being small, static, immutable, and eternal it became enormous, expanding,

---

[2] As noted by Peebles, it is also true that, until this conclusion was reached, a long series of events had to occur: studies, observations, and previous results of capital importance, that led to the final great confirmation of the CMB.

and endowed with an origin out of 'nothing' (1st revolution). Then, we also learned that its expansion accelerates without stopping, thus creating an absolute mystery where almost complete certainty had reigned before (2nd revolution). For me, looking at them in perspective, these are unquestionably two impressive revolutions. Indeed, two of the greatest discoveries in human history.

Bibliography

1. Leakey, M.D. 1971. *Olduvai Gorge: Excavations in Beds I & II 1960–1963*. Cambridge: Cambridge University Press.
2. Rogers, A.R., N.S. Harris, and A.A. Achenbach. 2020. Neanderthal-Denisovan Ancestors Interbred with a Distantly Related Hominin. *Science Advances* 6: 5483.
3. Henshilwood, C.S., F. d'Errico, K.L. van Niekerk, L. Dayet, A. Queffelec, and L. Pollarolo. 2018. An Abstract Drawing from the 73,000-Year-Old Levels at Blombos Cave, South Africa. *Nature* 562: 115–118. https://doi.org/10.1038/s41586-018-0514-3.
4. Adam Brumm, A., A.A. Adhi Agus Oktaviana, B. Basran Burhan, B. Budianto Hakim, R. Rustan Lebe, J. Jian-xin Zhao, P.H. Sulistyarto, M. Ririmasse, S. Adhityatama, I. Sumantri, and M. Aubert. 2021. Oldest Cave Art Found in Sulawesi. *Science Advances* 7 (3): abd4648. https://doi.org/10.1126/sciadv.abd4648.
5. Folkerts, M., E. Knobloch, and K. Reich. 2018. *Maß, Zahl und Gewicht, Mathematik als Schlüssel zu Weltverständnis und Weltbeherrschung*, Hrsg. v. M. Folkerts, E. Knobloch, and K. Reich, 13–40. Weinheim: Jochen Bär, Marcus Müller.
6. Ptolomeo. 1515. *Almagestum*. Universidad de Viena. http://www.univie.ac.at/hwastro/rare/1515_ptolemae.htm.
7. Halley, E. 1714–16. An Account of Several Nebulae or Lucid Spots Like Clouds, Lately Discovered Among the Fixt Stars by Help of the Telescope. *Philosophical Transactions* XXXIX: 390–392.
8. Messier, Ch. 1781. Catalogue des Nébuleuses et des Amas d'Étoiles. *Connaissance des Temps ou des Mouvements Célestes* 1784: 227–267.

9. Herschel, W., and C. Herschel. 1786. Catalogue of One Thousand New Nebulae and Clusters of Stars. *Philosophical Transactions* 76. Royal Society GB.

10. *Ernst Öpik*. Wikipedia. https://es.wikipedia.org/wiki/Ernst_%C3%96pik.

11. Hubble, E.P. 1937. *The Observational Approach to Cosmology*. Oxford.

12. Bartusiak, M. 2010. *The Day We Found the Universe*. Random House Digital.

13. The Stanford Encyclopedia of Philosophy, Library of Congress Catalog, 2019. ISSN 1.095-5.054. https://plato.stanford.edu/index.html.

14. Savage, C.W. (ed.). 1990. Preface. In *Scientific Theories, Minnesota Studies in the Philosophy of Science*, vol. 14. Minneapolis: University of Minnesota Press.

15. Richardson, A., and T. Übel (eds.). 2007. The Structure of Scientific Theories in Logical Empiricism. In *The Cambridge Companion to Logical Empiricism*, 136–162. Cambridge: Cambridge University Press.

16. Morris, C.W. 1938. *Foundations of the Theory of Signs. International Encyclopedia of Unified Science*. The University of Chicago Press. https://pure.mpg.de/rest/items/item_2364493/component/file_2364492/content.

17. Peirce, C.S. 1934–35. *Collected Papers*. Cambridge, MA: Harvard University Press.

18. Horgan, J. 2014. Opinion: Science Is Running Out of Things to Discover. *National Geographic*, Apr 4. https://www.nationalgeographic.com/travel/article/140409-nobel-prize-physics-aging-scientists-string-theory-inflation.

19. Woit, P. 2007. *Not Even Wrong: The Failure of String Theory and the Search for Unity in Physical Law*, Reprint ed. Basic Books.

20. Horgan, J. 2015. *The End of Science: Facing the Limits of Knowledge in the Twilight of the Scientific Age*, New ed. Basic Books.

21. Elizalde, E., and J. Gomis. 1978. The Groups of Poincaré and Galilei in Arbitrary Dimensional Spaces. *Journal of Mathematical Physics* 19: 1790; Elizalde, E. 1978. Poincare is a Subgroup of Galilei in One Space Dimension More. *Journal of Mathematical Physics* 19: 526.

22. Leavitt, H.S., and E.C. Pickering. 1912. Periods of 25 Variable Stars in the Small Magellanic Cloud. *Harvard College Observatory Circulars* 173: 1–3; Leavitt, H.S. 1908. 1777 Variables in the Magellanic Clouds. *Annals of Harvard College Observatory* LX (IV): 60, 87–110.

23. Byers, N., and G. Williams (eds.). 2006. *Out of the Shadows: Contributions of Twentieth-Century Women to Physics*. Cambridge: Cambridge University Press; Lightman, A. 2010. *The Discoveries: Great Breakthroughs in 20th-Century Science*. Canada: Knopf.

24. *Henrietta Swan Leavitt*. Wikipedia. https://en.wikipedia.org/wiki/Henrietta_Swan_Leavitt.

25. Bernstein, J. 2005. Book Review: George Johnson's Miss Leavitt's Stars. *Los Angeles Times*, July 17.

26. *1912: Henrietta Leavitt Discovers the Distance Key*. http://cosmology.carnegiescience.edu/timeline/1912.

27. Tao, L., E. Spiegel, and O.M. Umurhan. (1998). *Stellar Oscillations*. APS Division of Fluid Dynamics Meeting Abstracts. Princeton Lesson on Radial Stellar Pulsation, with the Kappa and Epsilon Mechanisms. https://www.astro.prince ton.edu/~gk/A403/pulse.pdf.

28. LeBlanc, F. 2010. *An Introduction to Stellar Astrophysics*. Wiley.

29. Elizalde, E. 2012. Cosmological Constant and Dark Energy: Historical Insights. In *Open Questions in Cosmology*, ed. by G.J. Olmo, Chap. 1. Geneva: InTech Publishers. ISBN 978-953-51-0880-1. https://doi.org/10.5772/51697; Elizalde, E. 2012. L'Origen i el Futur de l'Univers. In *La Terra a l'Univers: Astronomia*, 186–194. Barcelona: Enciclopedia Catalana; Elizalde, E. Blog (in Catalan). https://www.enciclopedia.cat/divulcat/Emili-Elizalde.

30. Hoyt, W.G. 1980. *Vesto Melvin Slipher, Biographical Memoirs*. Washington: National Academy of Sciences. http://www.nasonline.org/publications/biogra phical-memoirs/memoir-pdfs/sliphervesto.pdf.

31. Slipher, V.M. 2012. The Radial Velocity of the Andromeda Nebula. *Lowell Observatory Bulletin* 2 (58): 56–57.

32. Slipher, V.M. 1914. The Discovery of Nebular Rotation. *Scientific American* 110: 501.

33. Hall, J.S. 1969. Slipher's Trail-Blazing Career, Vesto Slipher. *Arizona Daily*, Sun, November 9.

34. Shapley, H., and H.D. Curtis. 1921. The Scale of the Universe. *Bulletin of the National Research Council* 2: 171–217; Gott, J.R., III, M. Juric, D. Schlegel, F. Hoyle, M. Vogeley, M. Tegmark, N. Bahcall, and J. Brinkmann. 2005. A Map of the Universe. *The Astrophysical Journal* 624: 463–484; Shu, F. 1982. *The Physical Universe, An Introduction to Astronomy*, 286. Mill Valley, CA: University Science Books; Nemiroff, R., and J. Bonnell (Org.). Great Debates in Astronomy. http://apod.nasa.gov/diamond_jubilee/debate. html; Trimble, V. 1995. The 1920 Shapley-Curtis Discussion: Background, Issues, and Aftermath. *Publications of the Astronomical Society of the Pacific* 107: 1133–1144.

35. Wilson, E.O. 2017. *The Origins of Creativity*. Liveright.

36. Hubble's Famous M31 VAR! Plate. http://obs.carnegiescience.edu/PAST/ m31var.

37. Croswell, K. 2001. *The Universe at Midnight: Observations Illuminating the Cosmos*. Free Press.

38. Kant, I. 1755. *Allgemeine Naturgeschichte und Theorie des Himmels, nach Newtonischen Grundsatzen Abgehandelt*. Königsberg und Leipzig: Johann Friederich Petersen. http://books.google.com/books?id=zbFDAAAAcAAJ.

39. Way, M.J. 2018. Dismantling Hubble's Legacy? In *ASP Conference Series*. Astronomical Society of the Pacific. https://ntrs.nasa.gov/archive/nasa/casi. ntrs.nasa.gov/20150019759.pdf.

40. Herschel, W. 1785. On the Construction of the Heavens. *Philosophical Transactions of the Royal Society of London Series I* 75: 213.

41. Lundmark, K. 1927. *Studies of Anagalactic Nebulae—First Paper*. Stockholm: Almquist & Wiksell Chap. 1.

42. Öpik, E. 1922. An Estimate of the Distance of the Andromeda Nebula. *The Astrophysical Journal* 55: 406–410; Erratum. 1923. *The Astrophysical Journal* 57: 192.

43. Öpik, Ernst. 1916. The Densities of Visual Binary Stars. *The Astrophysical Journal* 44: 292–302.

44. Clark, Ronald W. 2007. *Einstein: The Life and Times*. William Morrow Ed.

45. Kuepper, H.-J. *List of Scientific Publications of Albert Einstein*. https://einstein-website.de/.

46. https://en.wikipedia.org/wiki/Albert_Einstein, Ref. 174. https://farm3.static.flickr.com/2687/4496554935_0b573db853_o.jpg.

47. Pais, A. 1982. *Subtle is the Lord: The Science and the Life of Albert Einstein*, 179–183. Oxford University Press.

48. Stachel, J., et al. (eds.). 2008. The Collected Papers of Albert Einstein. In *Einstein's Writings*, 1–10. Published 1987–2006. Princeton University Press.

49. Pais, A. 1982. *Subtle is the Lord: The Science and the Life of Albert Einstein*, 194–195. Oxford University Press.

50. Berenguer, M. 2018. *Aproximació històrica a les equacions de camp d'Einstein*. Treball de Pràctiques Externes. Universitat Autònoma de Barcelona, Setembre.

51. van Dongen, J. 2010. *Einstein's Unification*, 23. Cambridge University Press.

52. Einstein, A. 1918. Über Gravitationswellen. In *Sitzungsberichte der Königlich Preussischen Akademie der Wissenschaften Berlin, Part 1*, 154–167.

53. *Gravity Investigated with a Binary Pulsar*. Press Release: The 1993 Nobel Prize in Physics. Nobel Foundation.

54. Abbott, B.P., et al. (LIGO Scientific Collaboration and Virgo Collaboration). 2016. Observation of Gravitational Waves from a Binary Black Hole Merger. *Physical Review Letters* 116: 061102. arXiv:1602.03837.

55. Gravitational Waves: Ripples in the Fabric of Space-Time, LIGO—MIT. 2016. Scientists Make First Direct Detection of Gravitational Waves, Jennifer Chu. *MIT News*, February 11. http://news.mit.edu/2016/ligo-first-detection-gravitational-waves-0211; Ghosh, P. 2016. Einstein's Gravitational Waves 'Seen' from Black Holes. *BBC News*, February 11; Overbye, D. 2016. Gravitational Waves Detected, Confirming Einstein's Theory. *The New York Times*, February 11.

56. Kennefick, D. 2019. *No Shadow of a Doubt. The 1919 Eclipse That Confirmed Einstein's Theory of Relativity*. Princeton, Oxford: Princeton University Press.

57. Einstein, A. 1924. Quantentheorie des einatomigen idealen Gases. In *Sitzungsberichte der Preussischen Akademie der Wissenschaften, Physikalisch-Mathematische Klasse*, 261–267.

58. McGehan, F., and P. Caughey, 2001. Cornell, and Wieman Share 2001 Nobel Prize in Physics. October 9. https://web.archive.org/web/20070610080506/https://www.nist.gov/public_affairs/releases/n01-04.htm.

59. Einstein, A., B. Podolsky, and N. Rosen. 1935. Can Quantum-Mechanical Description of Physical Reality Be Considered Complete? *Physical Review* 47: 777–780.

60. Isaacson, W. 2007. *Einstein: His Life and Universe*. New York: Simon & Schuster.

61. Penrose, R. 2007. *The Road to Reality*. Vintage Books.

62. Fine, A. 2017. *The Einstein-Podolsky-Rosen Argument in Quantum Theory*. Stanford Encyclopedia of Philosophy. Metaphysics Research Lab, Stanford University.

63. Einstein, A. 1950. On the Generalized Theory of Gravitation. *Scientific American* CLXXXII: 13–17.

64. *Result of WordNet Search for Einstein*. The Trustees of Princeton University. http://wordnetweb.princeton.edu/perl/webwn?s=Einstein.

65. Wilczek, F. 2004. Total Relativity: Mach 2004. *Physics Today* 57 (4): 10. https://physicstoday.scitation.org. https://doi.org/10.1063/1.1752398.

66. Einstein, A. 1917. *Kosmologische Betrachtungen zur allgemeinen Relativitäts-theorie*, 142–152. Sitzungsberichte der Königlich Preußischen Akademie der Wissenschaften.

67. Suhendro, I. 2008. Biography of Karl Schwarzschild. *The Abraham Zelmanov Journal* 1; O'Connor, J.J., and E.F. Robertson. Karl Schwarzschild. In *MacTutor History of Mathematics Archive*. University of St Andrews. https://en.wikipedia.org/wiki/Karl_Schwarzschild.

68. Letter from K Schwarzschild to A Einstein Dated 22 December 1915. In *The Collected Papers of Albert Einstein*, vol. 8a, doc. #169 (Transcript of Schwarzschild's Letter to Einstein of 22 Dec 1915). Archived 2012-09-04 at the Wayback Machine.

69. Eisenstaedt, J. 1989. The Early Interpretation of the Schwarzschild Solution. In *Einstein and the History of General Relativity: Einstein Studies*, vol. 1, ed. D. Howard and J. Stachel, 213–234. Boston: Birkhauser.

70. O'Connor, J.J., and E.F. Robertson. Willem de Sitter. In *MacTutor History of Mathematics Archive*. University of St Andrews. https://en.wikipedia.org/wiki/Willem_de_Sitter.

71. de Sitter, W. 1916. On Einstein's Theory of Gravitation and Its Astronomical Consequences, First Paper. *Monthly Notices of the Royal Astronomical Society* 76: 699–728; de Sitter, W. 1916. On Einstein's Theory of Gravitation and Its Astronomical Consequences, Second Paper. *Monthly Notices of the Royal Astronomical Society* 77: 155–184; de Sitter, W. 1917. On Einstein's Theory of Gravitation and Its Astronomical Consequences, Third Paper. *Monthly Notices of the Royal Astronomical Society* 78: 3–28.

72. Hockey, T. 2009. *The Biographical Encyclopedia of Astronomers*, Springer Publishing.

73. Beenakker, C. *Friedmann papers*. Instituut Lorentz, Leiden University. http://www.lorentz.leidenuniv.nl/Friedmann/.

74. Davidson, P.A., et al. (eds.). 2011. *A Voyage Through Turbulence*. Cambridge University Press.

75. Tropp, E.A., V.Y. Frenkel, and A.D. Chernin. 2006. *The Final Year. Alexander A Friedmann: The Man Who Made the Universe Expand*. Cambridge University Press.

76. Wirtz, C. 1922. Einiges zur Statistik der Radialbewegungen von Spiralnebeln und Kugelsternhaufen. *Astronomische Nachrichten* 215: 349–354.

77. Wirtz, C. 1924. De Sitters Kosmologie und die Radialbewegungen der Spiralnebel. *Astronomische Nachrichten* 222: 21–26.

78. Kinney, A.C., D. Calzetti, R.C. Bohlin, K. McQuade, T. Storchi-Bergmann, and H.R. Schmitt. 1996. Template Ultraviolet Spectra to Near-Infrared Spectra of Star-Forming Galaxies and Their Application to K-Corrections. *The Astrophysical Journal* 467: 38–60; North, J.D. 1965. *The Measure of The Universe: A History of Modern Cosmology*. Oxford University Press.

79. Wirtz, C. 1918. Über die Bewegungen der Nebelflecke. Vierte Mitteilung. *Astronomische Nachrichten* 206: 109–112.

80. van den Bergh, S. 2011. *Discovery of the Expansion of the Universe*. arXiv:1108. 0709 [physics.hist-ph].

81. Wirtz, C. 1922. Notiz zur Radialbewegung der Spiralnebel. *Astronomische Nachrichten* 215: 451.

82. Wirtz, C. 1936. Ein literarischer Hinweis zur Radialbewegung der Spiralnebel. *Zeitschrift für Astrophysik* 11: 261.

83. https://dspace.mit.edu/discover;   https://dspace.mit.edu/handle/1721.1/ 10753;   https://dspace.mit.edu/bitstream/handle/1721.1/10753/36897534- MIT.pdf?sequence=2&isAllowed=y.

84. Lemaître, G. 1927. Un Univers homogène de masse constante et de rayon croissant rendant compte de la vitesse radiale des nébuleuses extra-galactiques. *Annales de la Société Scientifique de Bruxelles* 47: 49.

85. Strömberg, G. 1925. Analysis of Radial Velocities of Globular Clusters and Non-Galactic Nebulae. *The Astrophysical Journal* 61: 353.

86. Slipher, V. 1917. Nebulae. *Proceedings of the American Philosophical Society* 56: 403–409.

87. Eddington, A.S. 1923. *The Mathematical Theory of Relativity*. Cambridge: Cambridge University Press.

88. Hubble, E. 1929. A Relation Between Distance and Radial Velocity Among Extra-Galactic Nebulae. *Proceedings of the National academy of Sciences of the United States of America* 15: 168–173.

89. Peacock, J.A. 2013. Slipher, galaxies, and cosmological velocity fields. In *Proceedings of the "Origins of the Expanding Universe: 1912–1932"*, vol. 471, ed. by M.J. Way and D. Hunter. San Francisco, CA: Astronomical Society of the Pacific.

90. Hubble, E.P., and M.L. Humason. 1931. The Velocity-Distance Relation Among Extra-Galactic Nebulae. *The Astrophysical Journal* 74: 43.

91. Odenwald, S., and R. Fienberg. 1993. *Redshifts Reconsidered*. New York: Sky Pub Co.

92. Nussbaumer, H. 2014. Einstein's Conversion from his Static to an Expanding Universe. *European Physical Journal* 39: 37–62; Nussbaumer, H., and L. Bieri. 2009. *Discovering the Expanding Universe*. Cambridge University Press; Ferris, T. 1977. *The Red Limit: The Search for the Edge of the Universe*. Bantam Books.

93. O'Raifeartaigh, C. 2018. Investigating the Legend of Einstein's "Biggest Blunder". *Physics Today*, October 30. https://physicstoday.scitation.org. https://doi.org/10.1063/PT.6.3.20181030a/full/.

94. Einstein, A. 2011. *The Meaning of Relativity, Four Lectures Delivered at Princeton University, May, 1921*. EBook-No. 36276. Project Gutenberg.

95. Savant Refutes Theory of Exploding Universe—Mt. Wilson Astronomer Reports Results of Long Searching with 100-Inch Telescope. *The Los Angeles Times*, December 31, 1941; Harnisch, L. 2011. Hubble: No Evidence of 'Big Bang' Theory. *LA Daily Mirror*, December 31.

96. Sandage, A. 1989. Edwin Hubble 1889–1953. *The Journal of the Royal Astronomical Society of Canada* 83: 6.

97. Hubble, E.P. 1958. *The Realm of the Nebulae*. New York: Dover Pub. Inc. https://archive.org/details/TheRealmOfTheNebulae.

98. Eddington, A.S. 1930. On the Instability of Einstein's Spherical World. *Monthly Notices of the Royal Astronomical Society*, 668–688.

99. Lemaître, Abbé G. 1931. A Homogeneous Universe of Constant Mass and Increasing Radius Accounting for the Radial Velocity of Extra-Galactic Nebulae. *Monthly Notices of the Royal Astronomical Society* 91: 483–490.

100. Livio, M. 2011. Lost in Translation: Mystery of the Missing Text Solved. *Nature* 479: 171–173.

101. General Assembly. 2018. *Resolutions to be voted at the XXXth General Assembly*. International Astronomical Union. https://www.iau.org/news/announcements/detail/ann18029/.

102. Lemaître, G. 1950. L'expansion de l'Univers. *Annales d'Astrophysique* 13: 344.

103. Elizalde, E. 2019. Reasons in Favor of a Hubble-Lemaître-Slipher's (HLS) Law. *Symmetry* 11: 35.

104. Robertson, H.P. 1928. On Relativistic Cosmology. *Philosophical Magazine* 5: 835.

105. Stigler, S.M. 1980. Stigler's Law of Eponymy. *Transactions of the New York Academy of Sciences* 39: 147–158.

106. Lemaître, G. 1931. The Beginning of the World from the Point of View of Quantum Theory. *Nature* 127: 706.

107. Menzel, D. 1932. *A Blast of Giant Atom Created Our Universe*, 52. Popular Science, Bonnier Corporation.

108. Elizalde, E. 2018. "All That Matter … in One Big Bang …", & Other Cosmological Singularities. *Galaxies* 6: 25.

109. Lambert, D. 2020. Einstein and Lemaître: Two Friends, Two Cosmologies… http://inters.org/einstein-lemaitre.

110. Kragh, H. 1996. *Cosmology and Controversy*, 55. Princeton University Press; Kragh, H., and R.W. Smith. 2003. Who Discovered the Expanding Universe? *History of Science* xli. https://doi.org/10.1177/007327530304100202.

111. Pope Pio XII. 1951. *Ai Cardinali, ai Legati delle Nazioni Estere e ai Soci della Pontificia Accademia delle Scienze*, November 22. https://www.vatican.va.

112. Lambert, D. 1997. Monseigneur Georges Lemaître et le débat entre la cosmologie et la foi (à suivre). *Revue Théologique de Louvain* 28: 28–53.

113. Georges Lemaître: Who Was the Belgian Priest Who Discovered the Universe is Expanding? *The Independent*, July 17, 2018. https://www.independent.co.uk/news/science/georges-lemaitre-priest-universe-expanding-big-bang-hubble-space-cosmic-egg-astronomer-physics-a8449926.html.

114. Zwicky, F. 1933. Die Rotverschiebung von extragalaktischen Nebeln. *Helvetica Physica Acta* 6: 110–127.

115. Zwicky, F. 1937. Nebulae as Gravitational Lenses. *Physical Review* 51: 290.

116. Rubin, V.C., and W.K. Ford Jr. 1970. Rotation of the Andromeda Nebula from a Spectroscopic Survey of Emission Regions. *The Astrophysical Journal* 159: 379–403.

117. The XENON Dark Matter Project; http://xenon.astro.columbia.edu/; cordis.europa.eu/project/rcn/101085_en.html; http://www.sciencemag.org/news/2014/07/two-big-dark-matter-experiments-gain-us-support.

118. Hoyle, F. 1946. The Synthesis of the Elements from Hydrogen. *Monthly Notices of the Royal Astronomical Society* 106: 343–383.

119. Burbidge, E.M., G.R. Burbidge, W.A. Fowler, and F. Hoyle. 1957. Synthesis of the Elements in Stars. *Reviews of Modern Physics* 29: 547.

120. Hoyle, F. 1954. On Nuclear Reactions Occurring in Very Hot STARS. I. The Synthesis of Elements from Carbon to Nickel. *The Astrophysical Journal Supplement Series* 1: 121–146.

121. O'Raifeartaigh, C., B. McCann, W. Nahm, and S. Mitton. *Einstein's Steady-State Model of the Universe*. arxiv.org/vc/arxiv/papers/1402/1402.0132v1.pdf.

122. Tolman, R.C. 1934. *Relativity, Thermodynamics, and Cosmology*. Oxford: Oxford University Press.

123. Eddington, A.S. 2014. The Nature of the Physical World. In *Gifford Lectures of 1927. An Annotated Edition of H.G. Callaway*. Newcastle: Cambridge Scholars Pub.

124. Gribbin, J. 2005. Stardust Memories. *The Independent*, June 17.

125. Guth, A.H. 1981. Inflationary Universe: A Possible Solution to the Horizon and Flatness Problems. *Physical Review D* 23: 347.

126. Alpher, R.A., H. Bethe, and G. Gamow. 1948. The Origin of Chemical Elements. *Physical Review* 73: 803. The humorous inclusion of Bethe's name in the article is explained here: https://en.wikipedia.org/wiki/Alpher%E2%80%93Bethe%E2%80%93Gamow_paper.

127. Gamow, G. 1948. The Evolution of the Universe. *Nature* 162: 680–682.

128. Gamow, G. 1953. Expanding Universe and the Origin of Galaxies. *Kongelige Danske Videnskabernes Selskab* 39 (27): 1–15.

129. Wilson, R.W. 1978. The Cosmic Microwave Background Radiation. *Nobel Lecture*, 8-12-1978. http://nobelprize.org/nobel_prizes/physics/laurea tes/1978/wilson-lecture.pdf.

130. Dunham, T., Jr., and W.S. Adams. 1937. *Publications of the American Astronomical Society* 9: 5.

131. Gamow, G. 1948. The Origin of Elements and the Separation of Galaxies. *Physical Review* 74: 505; Gamow, G. 1948. The Evolution of the Universe. *Nature* 162: 680; Alpher, R.A., and R. Herman. 1948. On the Relative Abundance of the Elements. *Physical Review* 74: 1577; Gamow, G. 1947. *One, Two, Three... Infinity.* Viking Press, revised 1961; Dover P., 1974.

132. Preskill, J.P. 1979. Cosmological Production of Superheavy Magnetic Monopoles. *Physical Review Letters* 43: 1365.

133. Guth, A.H., and S.-H.H. Tye. 1980. Phase Transitions and Magnetic Monopole Production in the Very Early Universe. *Physical Review Letters* 44: 631; Erratum. 1980. *Physical Review Letters* 44: 963.

134. https://physicstoday.scitation.orghttps://physicstoday.scitation.orghttps:// www.cs.mcgill.ca/~rwest/wikispeedia/wpcd/wp/c/Cosmic_inflation.htm.

135. Linde, A. 1982. A New Inflationary Universe Scenario: A Possible Solution of the Horizon, Flatness, Homogeneity, Isotropy and Primordial Monopole Problems. *Physics Letters B* 108: 389.

136. Albrecht, A., and P.J. Steinhardt. 1982. Cosmology for Grand Unified Theories with Radiatively Induced Symmetry Breaking. *Physical Review Letters* 48: 1220.

137. Starobinsky, A.A. 1979. Spectrum of Relict Gravitational Radiation and the Early State of the Universe. *Journal of Experimental and Theoretical Physics Letters* 30: 682; Spectrum of Relict Gravitational Radiation and the Early State of the Universe. 1979. *Pis'ma v Zhurnal Èksperimental'noi i Teoreticheskoi Fiziki (Soviet Journal of Experimental and Theoretical Physics Letters)* 30: 719.

138. Starobinsky, A.A. 1980. A New Type of Isotropic Cosmological Models Without Singularity. *Physics Letters B* 91: 99–102.

139. Calcagni, G. 2017. *Classical and Quantum Cosmology.* Cham: Springer; Huterer, D. 2012. *Big Bang Theory: The Three Pillars.* Ann Arbor, MI: University of Michigan. http://www-personal.umich.edu/~huterer/EPO/ HOUSTON_2012/three_pillars_houston.pdf; Ellis, J., J.S. Hagelin, D.V. Nanopoulos, K. Olive, and M. Srednicki. 1984. Supersymmetric Relics from the Big Bang. *Nuclear Physics B* 238: 453–476; Gasperini, M., and G. Veneziano. 1993. Pre-Big-Bang in String Cosmology. *Astroparticle Physics* 1: 317–339; The Hot Big Bang Model. http://www.damtp.cam.ac.uk/research/ gr/public/bb_home.html; The History of the Universe. http://www.astro.cor nell.edu/academics/courses/astro201; The Standard Hot Big Bang Model of the Universe. https://faraday.physics.utoronto.ca/PVB/Harrison/GenRel/Big BangModel.html; Cosmology: The Hot Big Bang. https://www.ras.org.uk/pub lications/other-publications/2040-cosmology-big-bang; Big Bang Cosmology. https://map.gsfc.nasa.gov/universe/bb_theory.html; The Big Bang Model of

the Universe. http://astronomy.swin.edu.au/~gmackie/BigBang/universe.htm; The Big Bang Model. http://spiff.rit.edu/classes/phys240/lectures/bb/bb.html.

140. Weinberg, S. 1993. *The First Three Minutes: A Modem View of the Origin of the Universe*, 2nd ed. New York, NY: Basic Books; 1st ed. New York, NY: Basic Books, 1977.

141. *The History of the Universe.* http://www.astro.cornell.edu/academics/courses/astro201.

142. *The Hot Big Bang Model.* http://www.damtp.cam.ac.uk/research/gr/public/bb_home.html.

143. Peebles, P.J.E., D.N. Schramm, E.L. Turner, and R.G. Kron. 1991. The Case for the Relativistic Hot Big Bang Cosmology. *Nature* 352: 769–776.

144. Khoury, J., B.A. Ovrut, P.J. Steinhardt, and N. Turok. 2001. Ekpyrotic Universe: Colliding Branes and the Origin of the Hot Big Bang. *Physical Review D* 64: 123522; Khoury, J., B.A. Ovrut, N. Seiberg, P.J. Steinhardt, and N. Turok. 2002. From Big Crunch to Big Bang. *Physical Review D* 65: 086007.

145. Berger, A.L. (ed.). 1983. *The Big Bang and Georges Lemaître: Proceedings of a Symposium in Honor of G. Lemaître Fifty Years After His Initiation of Big-Bang Cosmology*, 9, Louvain-Ia-Neuve, Belgium, October 10–13, 1983. Berlin/Heidelberg: Springer.

146. Doyle, B. *The Growth of Order in the Universe.* The Information Philosopher. http://www.informationphilosopher.com/solutions/scientists/layzer/growth_of_order/.

147. Barbour, J., T. Koslowski, and F. Mercati. 2014. Identification of a Gravitational Arrow of Time. *Physical Review Letters* 113: 181101.

148. Aguirre, A.N. 2000. The Cosmic Background Radiation in a Cold Big Bang. *The Astrophysical Journal* 533: 1; Aguirre, A. 2001. Cold Big-Bang Cosmology as a Counterexample to Several Anthropic Arguments. *Physical Review D* 64: 083508.

149. Wetterich, C. 2014. Hot Big Bang Or Slow Freeze? *Physics Letters B* 736: 506–514; Ashtekar, A., T. Pawlowski, and P. Singh. 2006. Quantum Nature of the Big Bang: Improved Dynamics. *Physical Review D* 74: 084003; Murphy, G.L. 1973. Big-Bang Model Without Singularities. *Physical Review D* 8: 4231. When the End Is Just the Beginning: Exploring Cosmic Cycles. http://www.pbs.org/wgbh/nova/blogs/physics/2014/08/when-the-end-is-just-the-beginning-exploring-cosmic-cycles/.

150. Lifshitz, E.M., and I.M. Khalatnikov. 1963. Problems of relativistic cosmology, *Soviet Physics Uspekhi* 6: 495–522; Lifshitz, E.M., and I.M. Khalatnikov. 1963. Investigations in Relativistic Cosmology. *Advances in Physics* 12: 185–249; Khalatnikov, I.M., and E.M. Lifshitz. 1970. General Cosmological Solution of the Gravitational Equations with a Singularity in Time. *Physical Review Letters* 24: 76; Belinsky, V.A., E.M. Lifshitz, and I.M. Khalatnikov. 1970. An Oscillatory Mode of Approach to Singularities in Relativistic Cosmology. *Uspekhi Fizicheskikh Nauk* 102: 463–500; Belinskii, V.A., E.M. Lifshitz, and I.M.

Khalatnikov. 1971. Oscillatory Approach to the Singular Point in Relativistic Cosmology. *Soviet Physics Uspekhi* 13: 745.

151. Misner, C.W. 1969. Mixmaster Universe. *Physical Review Letters* 22: 1071.

152. Penrose, R. 1965. Gravitational Collapse and Space-Time Singularities. *Physical Review Letters* 14: 57.

153. Hawking, S., and G.F.R. Ellis. 1973. *The Large Scale Structure of Space-Time*. Cambridge: Cambridge University Press; Wald, R.M. 1984. *General Relativity*. Chicago: University of Chicago Press; Geroch, R. 1968. Local Characterization of Singularities in General Relativity. *Annals of Physics* 48: 526; Garfinkle, D., and J.M.M. Senovilla. 2015. The 1965 Penrose Singularity Theorem. *Class. Quantum Gravity* 32: 124008; Hawking, S. The Beginning of Time. Lecture. http://homepages.wmich.edu/~korista/hawking-time.html.

154. Urban, P. (ed.). 1975. Electromagnetic Interactions and Field Theory. In *Proceedings of the XIV International Universitätswochen für Kernphysik*, Schladming, Austria, February 24–March 7, 1975. Berlín: Springer.

155. Elizalde, E. 2017. From the creation of particles in the vacuum by an accelerated observer to space-time thermodynamics. *Journal of Physics. A: Mathematical and Theoretical* 50: 041001. Viewpoint, by invitation.

156. Elizalde, E. 2012. Ten Physical Applications of Spectral Zeta Functions, 2nd ed. In *Lecture Notes in Physics*, vol. 855. Berlin: Springer-Verlag; Elizalde, E., S.D. Odintsov, A. Romeo, A.A. Bytsenko, and S. Zerbini. 1994. *Zeta Regularization Techniques with Applications*. Singapore: World Scientific.

157. Hawking, S. 1975. Particle Creation by Black Holes. *Communications in Mathematical Physics* 43: 199; Erratum. 1976. *Communications in Mathematical Physics* 46: 20.

158. Davies, P.C.W. 1975. Scalar Production in Schwarzschild and Rindler Metrics. *Journal of Physics A: Mathematical and General* 8: 609.

159. Unruh, W.G. 1976. Notes on Black-Hole Evaporation. *Physical Review D* 14: 870.

160. Wald, R.M. 1993. Black Hole Entropy is the Noether Charge. *Physical Review D* 48: R3427.

161. Verlinde, E.P. 2011. On the Origin of Gravity and the Laws of Newton. *Journal of High Energy Physics JHEP* 04: 029; Padmanabhan, T. 2010. Thermodynamical Aspects of Gravity: New Insights. *Reports on Progress in Physics* 73: 046901.

162. Jacobson, T. 1995. Thermodynamics of Spacetime: The Einstein Equation of State. *Physical Review Letters* 75: 1260; Elizalde, E., and P.J. Silva. 2008. f(R) Gravity Equation of State. *Physical Review D* 78: 061501.

163. Bekenstein, J.D. 1973. Black Holes and Entropy. *Physical Review D* 7: 2333.

164. https://dspace.mit.edu/bitstream/handle/1721.1/10753/36897534-MIT.pdf%3Fsequence%3D2%26isAllowed%3Dyhttps://en.wikipedia.org/wiki/Micro_black_hole; https://home.cern/science/physics/extradimensions-gravitons-and-tiny-black-holes.

165. Hawking, S.W., and T. Hertog. 2018. A Smooth Exit from Eternal Inflation. *Journal of High Energy Physics* 04: 147.

166. Haco, S., S.W. Hawking, M.J. Perry, and A. Strominger. 2018. Black Hole Entropy and Soft Hair. *Journal of High Energy Physics* 12: 98.

167. Borde, A., and A. Vilenkin. 1993. Eternal Inflation and the Initial Singularity. *Physical Review Letters* 72: 3305.

168. Borde, A., A. Guth, and A. Vilenkin. 2003. Inflationary Spacetimes Are Incomplete in Past Directions. *Physical Review Letters* 90: 151301.

169. *Correct Interpretation of the Borde Guth Vilenkin (BGV) Theorem?* https://physics.stackexchange.com/questions/308325/correct-interpretation-of-the-borde-guth-vilenkin-bgv-theorem; *Let the Universe Be the Universe*. http://www.preposterousuniverse.com/blog/2012/09/25/let-the-universe-be-the-universe/; *The Beginning of the Universe*. http://inference-review.com/article/the-beginning-ofthe-universe; *Borde-Guth-Vilenkin Singularity Theorem*. http://creationwiki.org/Borde-Guth-Vilenkin_singularity_theorem. *Borde, Guth, and Vilenkin's Past-Finite Universe*. https://debunkingwlc.wordpress.com/2010/07/14/borde-guth-vilenkin/; *The Borde-Guth-Vilenkin Theorem, and More, on the "My Good Friend" Meme*. https://evolutionnews.org/2014/08/the_borde-guth/; *Did the Universe Begin? III: BGV Theorem*. http://www.wall.org/~aron/blog/did-theuniverse-begin-iii-bgv-theorem/.

170. Steinhardt, P.J., and N. Turok. 2002. A Cyclic Model of the Universe. *Science* 296: 1436–1439.

171. Krauss, L.M. 2012. *A Universe from Nothing: Why There Is Something Rather Than Nothing*. Hong Kong: Free Press.

172. Kohli, I.S. 2014. *Comments On: A Universe from Nothing*. arXiv:1405.6091, v6.

173. Krauss, L.M, and F. Wilczek. 2014. *From B-Modes to Quantum Gravity and Unification of Forces*. arXiv:1404.0634v3; Krauss, L.M, and F. Wilczek. 2014. Using Cosmology to Establish the Quantization of Gravity. *Physical Review D* 89: 047501.

174. Riess, A.G., et al. 1998. Observational Evidence from Supernovae for an Accelerating Universe and a Cosmological Constant. *The Astronomical Journal* 116: 1009–1038. arXiv:astro-ph/9805201; Perlmutter, S., et al. 1999. The Supernova Cosmology Project, Measurements of Ω and Λ from 42 High-Redshift Supernovae. *The Astrophysical Journal* 517: 565–586.

175. Blumenthal, G.R., S.M. Faber, J.R. Primack, and M.J. Rees. 1984. Formation of Galaxies and Large-Scale Structure with Cold Dark Matter. *Nature* 311: 517.

176. Silk, J., and N. Vittorio. 1987. Does Lower Omega Allow a Resolution of the Large-Scale Structure Problem? *The Astrophysical Journal* 317: 564.

177. Efstathiou, G., W. J. Sutherland, and S.J. Maddox. 1990. The Cosmological Constant and Cold Dark Matter. *Nature* 348: 705.

178. Kofman, L.A., and A.A. Starobinsky. 1985. *Pis'ma v Astronomicheskii Zhurnal* 11: 643. [English Translation in *Soviet Astronomy Letters* 11: 271 (1985)].

179. Górski, K.M., J. Silk, and N. Vittorio. 1992. Cold Dark Matter Confronts the Cosmic Microwave Background: Large-Angular-Scale Anisotropies in $\Omega_0+\lambda=1$ Models. *Physical Review Letters* 68: 733.

180. Peebles, P.J.E., and B. Ratra. 2003. The Cosmological Constant and Dark Energy. *Reviews of Modern Physics* 75: 559–606.

181. Weinberg, S. 2008. *Cosmology*. Oxford: Oxford University Press.

182. Gaztanaga, E. https://darkcosmos.com/home/f/isw-integrated-sachs%E2%80%93wolfe-effect. Ibid., Inside a Black Hole: the illusion of a Big Bang, *MNRAS*, to appear, hal-03106344v4.

183. Elizalde, E. 2009. El efecto Casimir. In *Investigación y Ciencia 3/2009*.

184. Weinberg, S. 1989. The Cosmological Constant Problem. *Reviews of Modern Physics* 61: 1; Weinberg, S. 1993. Dreams of a Final Theory: *The Search for the Fundamental Laws of Nature*. Vintage Press.

185. Sotiriou, T.P., and V. Faraoni. 2010. f(R) Theories of Gravity. *Reviews of Modern Physics* 82: 451.

186. Cognola, G., E. Elizalde, S. Nojiri, S.D. Odintsov, L. Sebastiani, and S. Zerbini. 2008. A Class of Viable Modified *f(R)* Gravities Describing Inflation and the Onset of Accelerated Expansion. *Physical Review D* 77: 046009.

187. Elizalde, E., S. Nojiri, S.D. Odintsov, L. Sebastiani, and S. Zerbini. 2011. Non-Singular Exponential Gravity: A Simple Theory for Early- and Late-Time Accelerated Expansion. *Physical Review D* 83: 086006.

188. Nojiri, S., S.D. Odintsov, and S. Tsujikawa. 2005. Properties of Singularities in (Phantom) Dark Energy Universe. *Physical Review D* 71: 063004.

189. Elizalde, E., S. Nojiri, and S.D. Odintsov. 2004. Late-Time Cosmology in (Phantom) Scalar-Tensor Theory: Dark Energy and the Cosmic Speed-Up. *Physical Review D* 70: 043539.

190. Ratra, B., and P.J.E. Peebles. 1988. Cosmology with a Time-Variable Cosmological 'Constant'. *The Astrophysical Journal* 325: L17; Ratra, B., and P.J.E. Peebles. 1988. Cosmological Consequences of a Rolling Homogeneous Scalar Field. *Physical Review D* 37: 3406; Ratra, B., and P.J.E. Peebles. 2003. The Cosmological Constant and Dark Energy. *Reviews of Modern Physics* 75: 559–606.

191. Thomas, K.D. 2015. *General Relativity's Influence and Mysteries*. Princeton: Institute for Advanced Study.

192. Charitos, P. 2016. Interview with James Peebles. *CERN EP Newsletter*. https://ep-news.web.cern.ch/content/interview-james-peebles.

193. de Lapparent, V., M.J. Geller, and J.P. Huchra. 1986. A Slice of the Universe. *The Astrophysics Journal Letters to the Editor* 302: L1–L5.

194. Peebles, P.J.E. 1971. *Physical Cosmology*. Princeton: Princeton University Press; Peebles, P.J.E. 1980. *The Large-Scale Structure of the Universe*. Princeton: Princeton University Press.

195. Peebles, P.J.E. 1993. *Principles of Physical Cosmology*. Princeton: Princeton University Press.

196. Caldwell, R.R. 2002. A Phantom Menace? Cosmological Consequences of a Dark Energy Component with Super-Negative Equation of State. *Physics Letters B* 545: 23–29; Caldwell, R.R., M. Kamionkowski, and N.N. Weinberg. 2003. Phantom Energy: Dark Energy with $w < -1$ Causes a Cosmic Doomsday. *Physical Review Letters* 91: 071301; Barrow, J.D. 2004. Sudden Future Singularities. *Class. Quantum Gravity* 21: L79; Barrow, J.D. 2004. More General Sudden Singularities. *Class. Quantum Gravity* 21: 5619; Bouhmadi-López, M., P.F. González-Díaz, and P. Martín-Moruno. 2008. Worse Than a Big Rip? *Physics Letters B* 659: 1–5.

197. Hinshaw, G., and NASA WMAP (eds.). 2006. *What Is the Universe Made of?*. https://www.esa.int/Science_Exploration/Space_Science/Extreme_space/What_is_the_Universe_made_of.

198. Riess, A.G., S. Casertano, W. Yuan, L.M. Macri, and D. Scolnic. 2019. Large Magellanic Cloud Cepheid Standards Provide a *1%* Foundation for the Determination of the Hubble Constant and Stronger Evidence for Physics beyond ΛCDM. *The Astrophysical Journal* 876: 85.

199. Brian Keating's Quest for the Origin of the Universe, The Joy of X, podcast by Steven Strogatz. https://www.quantamagazine.org/brian-keatings-quest-for-the-origin-ofthe-universe-20200331/.

200. Baum, L., and P.H. Frampton. 2008. Entropy of Contracting Universe in Cyclic Cosmology. *Modern Physics Letters A* 23: 33–36.

Index

© The Editor(s) (if applicable) and The Author(s), under exclusive
license to Springer Nature Switzerland AG 2021
E. Elizalde, *The True Story of Modern Cosmology*,
https://doi.org/10.1007/978-3-030-80654-5

Printed in the United States
by Baker & Taylor Publisher Services